Methods in Molecular Biology

Series Editor
John M. Walker
School of Life and Medical Sciences
University of Hertfordshire
Hatfield, Hertfordshire, UK

For further volumes:
http://www.springer.com/series/7651

For over 35 years, biological scientists have come to rely on the research protocols and methodologies in the critically acclaimed *Methods in Molecular Biology* series. The series was the first to introduce the step-by-step protocols approach that has become the standard in all biomedical protocol publishing. Each protocol is provided in readily-reproducible step-by-step fashion, opening with an introductory overview, a list of the materials and reagents needed to complete the experiment, and followed by a detailed procedure that is supported with a helpful notes section offering tips and tricks of the trade as well as troubleshooting advice. These hallmark features were introduced by series editor Dr. John Walker and constitute the key ingredient in each and every volume of the *Methods in Molecular Biology* series. Tested and trusted, comprehensive and reliable, all protocols from the series are indexed in PubMed.

Phytophthora

Methods and Protocols

Edited by

Monica L. Gerth

School of Biological Sciences, Victoria University of Wellington, Wellington, New Zealand

Rosie E. Bradshaw

Bioprotection Aotearoa, School of Food Technology and Natural Sciences, Massey University, Palmerston North, New Zealand

Humana Press

Editors
Monica L. Gerth
School of Biological Sciences
Victoria University of Wellington
Wellington, New Zealand

Rosie E. Bradshaw
Bioprotection Aotearoa, School of Food Technology and Natural Sciences
Massey University
Palmerston North, New Zealand

ISSN 1064-3745 ISSN 1940-6029 (electronic)
Methods in Molecular Biology
ISBN 978-1-0716-4329-7 ISBN 978-1-0716-4330-3 (eBook)
https://doi.org/10.1007/978-1-0716-4330-3

Cover Illustration Caption: 3D projection, Phytophthora pluvialis grown in V8 broth. Provided by Dr. Claudia Meisrimler.

This Humana imprint is published by the registered company Springer Science+Business Media, LLC, part of Springer Nature.
The registered company address is: 1 New York Plaza, New York, NY 10004, U.S.A.

Preface

Phytophthora species are plant pathogens that cause devastating diseases in crops and natural ecosystems worldwide. The economic impact of these diseases, such as late blight in potatoes caused by *Phytophthora infestans*, results in billions of dollars in losses annually. Therefore, effective molecular tools are essential for understanding and managing these threats.

We (the editors, Monica and Rosie) both came to work on *Phytophthora* after the emergence of kauri dieback disease in Aotearoa, New Zealand. Monica's background was in molecular microbiology of bacteria, while Rosie previously focused on fungal genetics. But kauri are ancient and treasured trees, native to New Zealand, and like many NZ scientists, this inspired us to pivot our research to try to contribute to finding solutions.

We quickly learned two key lessons: (i) molecular techniques we once took for granted, such as transformation, are challenging to implement in *Phytophthora* and often require extensive optimization for different species, and (ii) the *Phytophthora* research community is remarkably generous in sharing their hard-won knowledge, tools, and resources.

In this volume, we have compiled some of this collective wisdom, covering essential molecular techniques and methods for studying *Phytophthora*. The book includes methodologies such as CRISPR-Cas genome editing, which opens new avenues for investigating pathogenicity and resistance. It also addresses the tools for detecting, visualizing, and quantifying disease, and more.

We extend our thanks to all the authors for their contributions to this book. We hope that the practical advice, insider knowledge, tips, and tricks shared here will assist others as much as they have helped us. This book aims to equip researchers with the tools needed to address the challenges posed by *Phytophthora*, fostering innovation and collaboration in the field of *Phytophthora* molecular biology.

Wellington, New Zealand — *Monica L. Gerth*
Palmerston North, New Zealand — *Rosie E. Bradshaw*

Contents

Contributors

AUDREY M. V. AH-FONG • *Department of Microbiology and Plant Pathology, University of California, Riverside, CA, USA*

CHRISTINA E. ANDRONIS • *Proteomics International, Nedlands, WA, Australia; The Centre for Crop and Disease Management, Curtin University, Bentley, WA, Australia*

PAUL R. J. BIRCH • *Division of Plant Sciences, School of Life Sciences, James Hutton Institute, University of Dundee, Dundee, UK; Cell and Molecular Sciences, James Hutton Institute, Dundee, UK*

PETRA C. BOEVINK • *Cell and Molecular Sciences, James Hutton Institute, Dundee, UK*

ROSIE E. BRADSHAW • *Bioprotection Aotearoa, School of Food Technology and Natural Sciences, Massey University, Palmerston North, New Zealand*

SCOTT BRINGANS • *Proteomics International, Nedlands, WA, Australia*

SABINE BRUMM • *Sainsbury Laboratory, University of Cambridge, Cambridge, UK*

NICHOLAS C. CAULDRON • *Department of Botany and Plant Pathology, Oregon State University, Corvallis, OR, USA*

NAN CHEN • *Sanya Institute of Nanjing Agricultural University, Department of Plant Pathology, Nanjing Agricultural University, Nanjing, Jiangsu, China*

DAVID E. L. COOKE • *The James Hutton Institute, Dundee, UK*

JO ANNE CROUCH • *United States Department of Agriculture, Agricultural Research Service, Foreign Disease/Weed Science Research Unit, Frederick, MD, USA*

SOPHIE ECCERSALL • *School of Biological Sciences, University of Canterbury, Christchurch, New Zealand*

ASHLEY GARRILL • *Biomolecular Interaction Centre, University of Canterbury, Christchurch, New Zealand; School of Biological Sciences, University of Canterbury, Christchurch, New Zealand*

MONICA L. GERTH • *School of Biological Sciences, Victoria University of Wellington, Wellington, New Zealand*

GLEN GROBEN • *United States Department of Agriculture, Agricultural Research Service, Foreign Disease/Weed Science Research Unit, Frederick, MD, USA; Oak Ridge Institute for Science and Education, Agricultural Research Service Research Participation Program, Oak Ridge, TN, USA*

NIKLAUS J. GRÜNWALD • *Horticultural Crops Disease and Pest Management Research Unit, United States Department of Agriculture-Agricultural Research Service, Corvallis, OR, USA*

MAX HAYHURST • *School of Biological Sciences, Victoria University of Wellington, Wellington, New Zealand*

HOWARD S. JUDELSON • *Department of Microbiology and Plant Pathology, University of California, Riverside, CA, USA*

BEATRIX KEILLOR • *The James Hutton Institute, Dundee, UK*

TREVOR S. LOO • *Bioprotection Aotearoa, School of Food Technology and Natural Sciences, Massey University, Palmerston North, New Zealand*

MATTHEW MACLEOD • *Sainsbury Laboratory, University of Cambridge, Cambridge, UK*

Claudia-Nicole Meisrimler • *School of Biological Sciences, University of Canterbury, Christchurch, New Zealand*
Carl S. Mendoza • *Department of Microbiology and Plant Pathology, University of California, Riverside, CA, USA*
Volker Nock • *Department of Electrical and Computer Engineering, University of Canterbury, Christchurch, New Zealand; The MacDiarmid Institute for Advanced Materials and Nanotechnology, Wellington, New Zealand; Biomolecular Interaction Centre, University of Canterbury, Christchurch, New Zealand*
Jade T. T. Palmer • *School of Biological Sciences, Victoria University of Wellington, Wellington, New Zealand*
Camilo H. Parada-Rojas • *Department of Botany and Plant Pathology, Oregon State University, Corvallis, OR, USA*
Min Qiu • *Sanya Institute of Nanjing Agricultural University, Department of Plant Pathology, Nanjing Agricultural University, Nanjing, Jiangsu, China; The Key Laboratory of Plant Immunity, Nanjing Agricultural University, Nanjing, Jiangsu, China; Key Laboratory of Soybean Disease and Pest Control (Ministry of Agriculture and Rural Affairs), Nanjing Agricultural University, Nanjing, Jiangsu, China*
Eva Randall • *The James Hutton Institute, Dundee, UK*
Hannah F. Robinson • *School of Biological Sciences, Victoria University of Wellington, Wellington, New Zealand*
Catalina Salgado-Salazar • *United States Department of Agriculture, Agriculture Research Service, Mycology and Nematology Genetic Diversity and Biology Laboratory, Beltsville, MD, USA*
Sebastian Schornack • *Sainsbury Laboratory, University of Cambridge, Cambridge, UK*
Martha A. Sudermann • *Department of Botany and Plant Pathology, Oregon State University, Corvallis, OR, USA*
Yiling Sun • *Department of Electrical and Computer Engineering, University of Canterbury, Christchurch, New Zealand; The MacDiarmid Institute for Advanced Materials and Nanotechnology, Wellington, New Zealand*
Kar-Chun Tan • *The Centre for Crop and Disease Management, Curtin University, Bentley, WA, Australia*
Mariana Tarallo • *Bioprotection Aotearoa, School of Food Technology and Natural Sciences, Massey University, Palmerston North, New Zealand*
Ayelen Tayagui • *Department of Electrical and Computer Engineering, University of Canterbury, Christchurch, New Zealand; The MacDiarmid Institute for Advanced Materials and Nanotechnology, Wellington, New Zealand; Biomolecular Interaction Centre, University of Canterbury, Christchurch, New Zealand*
Ramesh R. Vetukuri • *Department of Plant Breeding, Swedish University of Agricultural Sciences (SLU), Lomma, Sweden*
Jochem N. A. Vink • *School of Biological Sciences, Victoria University of Wellington, Wellington, New Zealand*
Nguyen N. T. Vo • *Department of Microbiology and Plant Pathology, University of California, Riverside, CA, USA*
Haixia Wang • *Division of Plant Sciences, School of Life Sciences, James Hutton Institute, University of Dundee, Dundee, UK*

YUANCHAO WANG • *Sanya Institute of Nanjing Agricultural University, Department of Plant Pathology, Nanjing Agricultural University, Nanjing, Jiangsu, China; The Key Laboratory of Plant Immunity, Nanjing Agricultural University, Nanjing, Jiangsu, China; Key Laboratory of Soybean Disease and Pest Control (Ministry of Agriculture and Rural Affairs), Nanjing Agricultural University, Nanjing, Jiangsu, China*
LYDIA R. J. WELSH • *Cell and Molecular Sciences Department, The James Hutton Institute, Invergowrie, Dundee, UK*
STEPHEN C. WHISSON • *Cell and Molecular Sciences Department, The James Hutton Institute, Invergowrie, Dundee, UK*

Chapter 1

Promoter Analysis and Dissection Using Reporter Genes, Comparative Genomics, and Gel Shift Assays in *Phytophthora*

Nguyen N. T. Vo and Howard S. Judelson

Abstract

Transcriptional regulation allows cells to execute developmental programs, maintain homeostasis, and respond to intra- and extracellular signals. Central to these processes are promoters, which in eukaryotes are sequences upstream of genes that bind transcription factors (TFs) and which recruit RNA polymerase to initiate mRNA synthesis. Valuable tools for studying promoters include reporter genes, which can be used to indicate when and where genes are activated. Moreover, functional regions within promoters (typically TF binding sites) can be identified by integrating reporter assays with promoter mutagenesis. These sites may also be revealed through comparative genomics, or by the DNA-protein binding procedure known as a gel shift or electrophoretic mobility shift assay (EMSA). The latter can also be used to test if a specific TF binds a DNA target or assess the binding kinetics or affinity of the complex. In this chapter, we describe procedures for expressing reporter genes in *Phytophthora*, assaying reporter activity, identifying functional sites within promoters, and testing purified TFs or proteins within nuclear extracts for DNA binding.

Key words Reporter gene, Oomycete, Transformation, Transcription factor, DNA-protein binding assay, Comparative genomics, Phylogenetic footprinting

1 Introduction

The genus *Phytophthora* is infamous for causing major losses on food and fiber crops, ornamentals, and plants in natural environments [1]. Critical to the growth and development of *Phytophthora* is transcriptional regulation, which largely involves the binding of transcription factors (TFs) to promoters, which are sequences upstream of the coding sequences of genes. Several TFs controlling metabolism or developmental processes such as sporulation have been identified in *Phytophthora* [2–8]. Learning more about TFs, their targets in promoters, and environmental cues that regulate promoters has the potential to lead to strategies for controlling oomycete diseases.

Monica L. Gerth and Rosie E. Bradshaw (eds.), *Phytophthora: Methods and Protocols*, Methods in Molecular Biology, vol. 2892, https://doi.org/10.1007/978-1-0716-4330-3_1,

Reporter genes are useful tools for monitoring in vivo the temporal and spatial patterns of gene expression and identifying the signaling events that drive cellular and developmental processes. A diversity of reporters useful in oomycetes have been developed, including fluorescent proteins such as green fluorescent protein from *Aequorea victoria* (GFP; [9]) and β-glucuronidase from the *uidA* gene of *E. coli* (GUS; [10]). A prerequisite for using these is a system for inserting DNA into the genome, i.e., transformation. When fused to a constitutive promoter, reporters have been shown to be useful for tracing the growth of *Phytophthora* within a plant host [9]. Insight into the biology of plant pathogens may also be revealed when the reporter is fused to a promoter sensitive to signals such as the plant cell wall, redox state, or stimuli regulating spore germination or effector secretion [11–13].

In combination with promoter mutagenesis, reporters can also be used to define regulatory sites within promoters. Past studies of *Phytophthora* promoters have indicated that TF binding sites typically reside within a few hundred bases of the transcription start site, which is consistent with the relatively small (400–500 nt) intergenic regions found within oomycete genomes [12, 14, 15]. Proteins that bind such sites can be investigated using gel shift assays, also known as electrophoretic mobility shift assays (EMSA; [16]). Such assays are often used to confirm the binding of a TF to a DNA site, but can also be employed to measure binding affinities and stoichiometries within the DNA-protein complex [17], or to delineate the sequences required for TF binding which are often semi-degenerate [18].

A common strategy for dissecting the structure of promoters begins by fusing sequences upstream of the translation start site with a reporter, which is then tested in vivo. *Cis*-regulatory elements are then identified by testing the effect of base changes or deletions. The function of many motifs can also be measured by fusing their sequences to a minimal promoter. The latter, also known as a core promoter, is a short sequence that allows the initiation complex to form but has been stripped of upstream regulatory sites [19]. Whether regions within a promoter bind a transcription factor (or other protein) can also be assessed in vitro through gel shift assays using crude nuclear extracts or purified TFs. Clues can also be obtained through phylogenetic footprinting [20]. This usually involves identifying conserved blocks within alignments of orthologous promoters, but it is also possible to search for over-represented motifs within unaligned regions. The latter approach can be extended beyond analyses of orthologs to studies of co-expressed genes mined from whole-genome transcriptome data [15, 21].

In this chapter, we describe approaches that have proved to be successful for studying *Phytophthora* promoters using reporter genes and gel shift assays. While our protocols focus on the use of the GUS gene and the potato pathogen *Phytophthora infestans,* the methods can be adapted to other reporters or species. We also discuss considerations that can be used when choosing a reporter, strategies for promoter mutagenesis, and ways in which comparative genomics and other bioinformatic tools can aid the analysis.

2 Materials

2.1 Constructing Reporter Plasmids

1. 37 °C incubators.
2. Microcentrifuge.
3. Transformation vector containing a selectable marker for drug resistance and a reporter gene containing a cloning site at its 5′ end. For GUS, a convenient vector is pNP-GUS [15], which contains the *nptII* marker (Fig. 1; Subheading 3, **step 1**). Promoter motifs can be tested for function using pNIFS-GUS, which contains a minimal promoter fused to GUS (Subheading 3, **step 6**).
4. PCR-amplified or synthesized promoter sequences.
5. Restriction enzymes for inserting promoter sequences into the reporter plasmids, such as *Xba*I and B*am*HI.
6. T4 DNA ligase.
7. Competent *E. coli* cells (e.g., DH5α).
8. Lysogeny Broth (LB): 1.0% NaCl, 1.0% tryptone, 0.5% yeast extract, with and without 1.5% agar.
9. TE buffer: 10 mM Tris pH 8.0, 1 mM EDTA.
10. Ampicillin.

2.2 Phytophthora Electroporation

1. Electroporator.
2. Electroporation cuvettes with a 4 mM gap.
3. Refrigerated centrifuge with swinging bucket rotor.
4. Light microscope.
5. Hemocytometer.
6. 15 μm nylon mesh.
7. Bent glass rod or disposable plastic spreader.
8. Regeneration media: Rye A (rye-sucrose) broth [22] is clarified by centrifugation at 5000× *g*, and then mannitol is added to 100 mM, $CaCl_2$ to 2.5 mM, and KCl to 1 mM. Other media may be preferred for species other than *P. infestans.*
9. 5 M LiCl, kept at 4 °C.
10. Modified Petri's solution (0.8 mM KCl, 0.25 mM $CaCl_2$, 1 mM KH_2PO_4, 1 mM $MgSO_4$), kept at 4 °C.

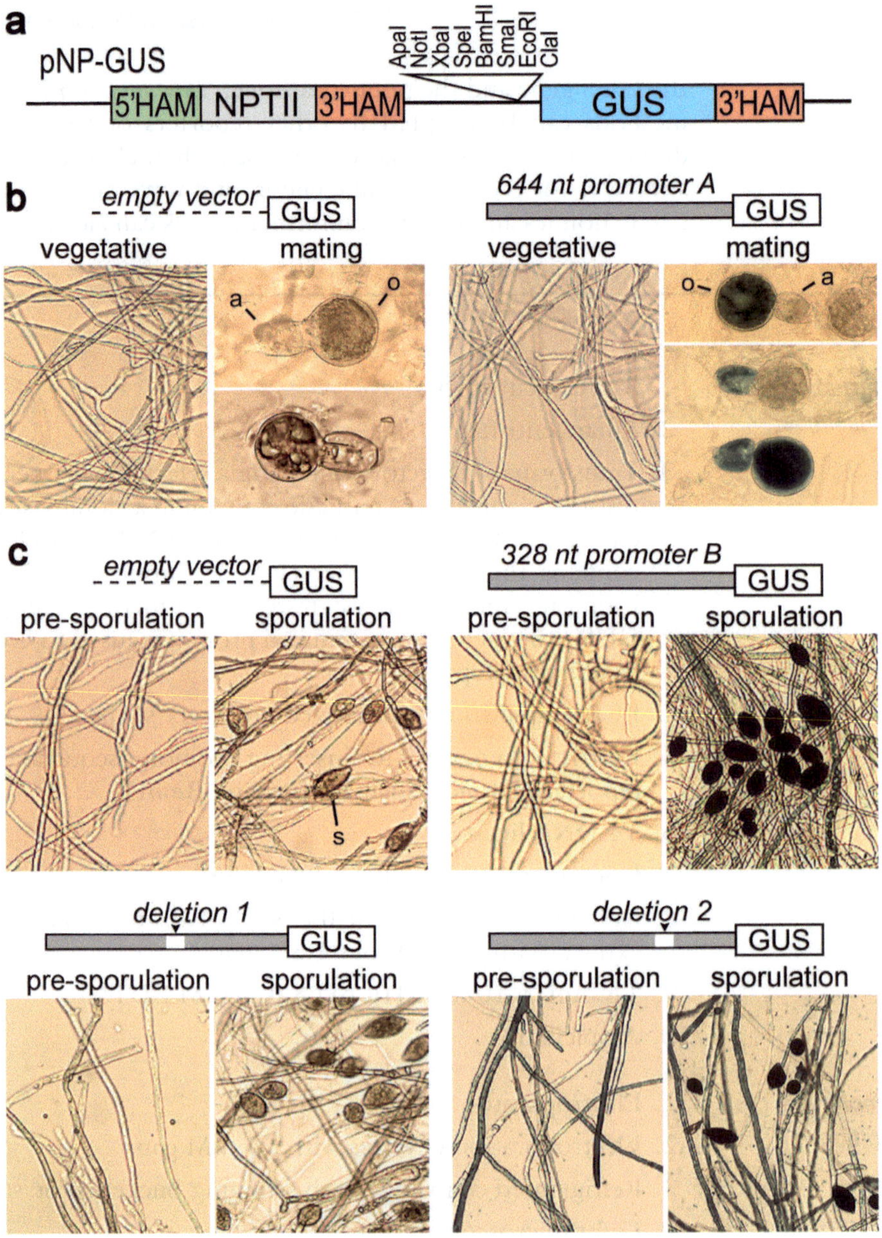

Fig. 1 Testing promoters using GUS. (**a**) Linearized schematic of pNP-GUS, which contains a promoterless GUS gene and *nptII* for resistance to G-418. (**b**) GUS driven by a 406-nt mating-induced promoter and detected by histochemical staining (blue color). No signal was detected using empty vector controls in vegetative hyphae, or in mating structures obtained by co-culture with an untransformed strain of the opposite mating type (oogonia, o, and antheridia, a). The oospore in the lower empty vector panel displays the darkening typical of maturing oospores, not the blue color from GUS. In contrast, the promoter-GUS fusion caused staining only in the sexual structures. Illustrated are expression in an antheridium (top right panel) or oogonium (middle right) in outcrossed pairings, and in both during selfing (lower right). (**c**) Analysis of a promoter induced during asexual sporulation. The intact 328-nt promoter is shown driving expression in sporangia (s) and nearby hyphae, but not in pre-sporulation cultures. The 328-nt region had been identified as sufficient for expression based on the analysis of sequential 5′ deletions starting from an 800-nt promoter. Erasure of a predicted motif (deletion 1) eliminated transcription, indicating that the block binds an activator. Loss of a second motif (deletion 2) resulted in constitutive expression, suggesting that region binds a repressor

11. Selective media: Rye A media with 1.5% agar and a drug for selecting transformants. For plasmids mentioned in this chapter, this is normally 8 μg/mL G-418.
12. Optional antibiotics to guard against contamination: Nystatin and/or Penicillin G.

2.3 Histochemical Staining for GUS

1. Equipment: Light microscope with slides and coverslips.
2. 37 °C incubator.
3. Staining solution: 50 mM $NaPO_4$, pH 7.0, 0.1% Triton X-100, 0.1% bromochloroindoyl-β-D-glucuronide (X-Gluc), 5 mM $K_3Fe(CN)_6$, 5 mM $K_4Fe(CN)_6$. X-Gluc is added from a 10% stock in dimethyl formamide. These should be stored in the dark at −20 °C.

2.4 Quantitative Assay for GUS

1. Equipment: Fluorometer, preferably one capable of handling 96-well plates.
2. Homogenization equipment: We normally use a mortar and pestle with liquid nitrogen. Hyphae or plant tissue may also be disrupted using a mechanical homogenizer such as a Polytron (Kinemetica).
3. Extraction buffer: 0.25 M sodium phosphate, pH 7.0, 10 mM ethylenediaminetetraacetic acid (EDTA), 50 mM β-mercaptoethanol, 0.5% sodium N-lauroylsarcosine, 0.5% Triton X-100.
4. Assay buffer: Extraction buffer plus 1 mM 4-methylumbelliferyl-β-D-glucuronide (4-MUG).
5. Stop buffer: 0.2 M sodium carbonate.
6. 4-Methylumbelliferone (4-MU) for making standard curve.

2.5 Nuclear Protein Extracts

1. Equipment: Bead-Beater (Biospec Products).
2. Refrigerated centrifuge.
3. Nuclear Buffer 1: 10 mM piperazine-N,N′-bis(2-ethanesulfonic acid) pH 7.0, 10 mM $MgCl_2$, 25% (vol/vol) glycerol, 10 mM β-mercaptoethanol, 0.1 mM phenylmethylsulfonyl fluoride (PMSF); add the PMSF just before use from a 200 mM stock solution in isopropanol. Cocktails of protease inhibitors may be substituted for the PMSF.
4. Nuclear Buffer 2: 10% glycerol, 15 mM HEPES, pH 7.9, 0.5 M KCl, 5 mM MgCl2, 0.5 mM EDTA, 1 mM dithiothreitol (DTT), 1% Triton X-100.
5. Nuclear Buffer 3: 15 mM HEPES pH 7.9, 100 mM KCl, 1 mM EDTA, 2 mM DTT, 15% glycerol, and 0.1 mM PMSF (added just before use).

6. 10% Triton X-100.
7. 3 M KCl.

2.6 Gel Shift Assay (EMSA)

1. Vertical electrophoresis unit including power supply and low fluorescence gel plates.
2. Instrument for scanning the gel for fluorescence (e.g., GE Typhoon or Azure Sapphire).
3. Oligonucleotides containing the target site. This would include a Cy5-labeled oligonucleotide containing the target (stored in the dark), and its unlabeled complementary strand.
4. Unlabeled oligonucleotides for use as a specific competitor (same as above but without the Cy5 label).
5. Unlabeled oligonucleotides for use as nonspecific competitor. This could be an unrelated sequence or the same as above but with the putative target altered, e.g., changing A to C, G to T, C to A, and T to G.
6. TE buffer for resuspending the oligos (10 mM Tris pH 8.0, 1 mM EDTA).
7. 50 mM KCl.
8. 100 mM $MgCl_2$.
9. 1 M Tris pH 7.9.
10. 10× Tris-glycine buffer (0.25 M Tris pH 8.3, 1.92 M glycine).
11. Gel running buffer: 100 mL 10× Tris-Glycine buffer, 62.5 mL 80% glycerol, dH_2O to 1 L.
12. Gels containing 5% acrylamide (29:1 acrylamide:bis), 5% glycerol, 10% (v/v) 10× Tris-glycine buffer. These are made by standard protocols using 10% ammonium persulfate (APS), and *N, N, N′, N′*-tetramethylethylenediamine (TEMED) for polymerization.
13. Binding Buffer A for DNA dilution: 20 mM HEPES pH 7.9, 1 mM DTT, 8 mM $MgCl_2$, 0.1 mM $ZnCl_2$. When using nuclear extracts, add 1 μg poly(dI-dC) per 15 μL.
14. Binding Buffer B for protein dilution: 186 μL dH_2O, 75 μL 80% glycerol, 3 μL 2 M Tris pH 7.9, 30 μL 1 M KCl, 3 μL 0.1 M DTT, 3 μL 5 mg/mL bovine serum albumin.
15. Orange Loading Dye: 10 mM Tris pH 7.9, 15% Orange G, 60% glycerol.

3 Methods

3.1 Design of Promoter Fused Reporter Gene Constructs

1. Select a suitable vector system. We typically use pNP-GUS, which contains multiple cloning sites upstream of a promoter-less β-glucuronidase (GUS) gene, along with the *nptII* marker which confers resistance to G-418. Other vectors can also be used, such as pGFPN [23] after deleting the *ham34* promoter (*see* **Note 1**).
2. Select a promoter region to fuse upstream of the reporter gene (*see* **Note 2**). This can be amplified by PCR using oligonucleotides containing restriction sites compatible with the vector. We normally add unique sites to the 5′ and 3′ ends of the promoter region to force directional cloning; check to ensure that those sites are not in the promoter region. An alternative is to have the sequence of interest synthesized; this is useful when mutagenized versions of the promoter are to be tested.
3. Digest the plasmid (e.g., pNP-GUS) and promoter insert with the appropriate restriction enzymes, and then purify the two DNAs using a commercial kit.
4. Mix the plasmid and insert in a 1:3 molar ratio and ligate. We normally perform ligations in a microcentrifuge tube containing 1 μL of 10× T4 DNA ligase buffer containing ATP, 1 μL (400 units) of T4 DNA ligase, 100 ng total of plasmid and insert DNA, plus water to 10 μL. Incubate overnight at 12 °C.
5. Mix 1–10 μL of the ligation reaction with 50 μL competent *E. coli* cells (e.g., DH5α), incubate on ice for 30 min, and heat at 42 °C for 90 s. Then, add 0.8 mL of LB broth and place in a shaking incubator at 37 °C for 1 h. Plate aliquots on LB agar containing 50–100 μg/mL ampicillin and incubate at 37 °C until colonies are observed, about 14–18 h.
6. Identify clones containing the desired insert by performing a small-scale plasmid prep followed by restriction digestion analysis and/or Sanger sequencing. Alternatively, identify clones with inserts on the original plates by PCR.
7. Prepare DNA for *Phytophthora* transformation using a scaled-up plasmid prep. In our hands, about 50 to 100 μg of DNA is typically adequate for obtaining enough transformants.

3.2 Phytophthora Transformation

Several strategies are available for transforming *Phytophthora* spp. and related oomycetes. With *P. infestans,* we have had the best success with protoplast-based and zoospore electroporation methods [23, 24]. While both appear equally satisfactory for promoter analyses, here we present an electroporation protocol.

1. Inoculate seven 15-cm rye-sucrose agar plates with ~10^4 sporangia. These are harvested from an 8- to 10-day-old 100-mm

culture plate by scraping with a sterile bent glass rod in 10 mL of sterile H_2O or Petri's solution. After counting with a hemocytometer, spread ~10^4 sporangia on each of the 15-cm plates with a sterile glass rod. Incubate in the dark at 18 °C (*see* **Note 3**).

2. After 8–9 days of growth, sporulation should have occurred. Harvest the sporangia by pouring 20 mL of ice-cold Petri's solution on each plate, followed by rubbing off the sporangia with the glass rod, and decanting the resulting milky solution into an empty 15-cm plate.
3. Induce zoosporogenesis by incubation at 10 °C for about 2 h (*see* **Note 3**).
4. To separate the zoospores from ungerminated or empty sporangia, filter through 15 μm nylon mesh into a 400-mL beaker. This and the following steps should be done on ice to retard zoospore encystment. Holders for nylon mesh are sold by Biodesign of New York. Alternatively, cell strainers that fit into 50 mL conical tubes can be used.
5. Transfer the zoospores into 50 mL conical tubes, add one-fiftieth of the volume of cold 5 M LiCl, and mix gently by inversion. Count the zoospore concentration using a hemocytometer.
6. Spin at 400× *g* for 5 min at 4 °C in a swinging bucket rotor.
7. Discard the liquid and resuspend the pellet by adding enough cold Petri's solution containing 0.1 M LiCl to set the concentration between 6×10^6 and 2.5×10^7 per mL.
8. For each reaction, gently mix 800 μL of zoospores with 30 μg DNA in a pre-chilled tube.
9. Pipette 800 μL of the zoospore-DNA mixture into a precooled 4 mm-gap cuvette. Close the cuvette, wipe the contacts dry with a tissue, and electroporate at 550 V, 1575 Ω, and 50 μF (*see* **Note 4**). A typical time constant is around 1.8 to 2.6 ms. If several cuvettes will be used with the same DNA, step 8 can be scaled up.
10. Immediately place the cuvette on ice and add 800 μL of regeneration media. Then, pipette the mixture into a 15 mL tube containing an additional 9 mL of regeneration media (*see* **Note 5**). Incubate the tube, lying on its side, at 18 °C for 20 h. Longer incubations may cause the resulting germinating cysts to stick together, making them difficult to spread on plates.
11. After the regeneration period, count the concentration of germinated cysts using a hemocytometer. Calculate the regeneration rate (germlings divided by the original number of zoospores per reaction); this normally ranges between 5 and 40%.

12. Spin the regenerated liquid cultures at 1000 × *g* for 5 min at room temperature. Decant the supernatant, retaining 1.6 mL of the liquid in the tube.
13. Resuspend the zoospores by gently pipetting, then spread 0.2 mL on each of eight 100-mm rye-sucrose agar plates containing 8 μg/mL G-418 (*see* **Note 6**). Incubate at 18 °C.
14. Colonies typically appear around day 8, and should be transferred to fresh plates containing G-418. It is prudent to check early so that colonies can be transferred before they overlap.

3.3 Identifying Transformants Expressing GUS by Histochemical Staining

1. Select several transformants for analysis. We typically aim to identify at least three GUS-expressing strains with a consistent expression pattern (*see* **Note** 7).
2. In a microcentrifuge tube, combine the GUS staining solution with the relevant *Phytophthora* sample which may include hyphae, sporangia, germinated cysts, oospores, plant tissue, etc. For spores, it may be useful to concentrate the tissue by gentle centrifugation. For hyphae, a small tuft of tissue scraped from the surface of a plant is usually sufficient. We normally perform the reaction in a volume of 50 μL, but add more staining solution as needed to keep the tissue submerged.
3. Incubate for 10 min to overnight; strong promoters may yield a useful signal (a blue color) after a few minutes while several hours may be needed for weak promoters. The signal will appear faster if incubation is performed at 37 °C, but good results can also be obtained at room temperature.
4. Visualize by light microscopy.

3.4 Quantitative GUS Assay

Precise measurements of GUS expression may sometimes be required. For example, promoter mutagenesis studies (Subheading 3, **step 5**) may result in a quantitative increase or decrease in expression not detected easily by histochemical staining, which is only semi-quantitative. Options for accurate measurements of the GUS reporter include enzymatic assays with colorimetric or fluorescent substrates, immunoblots, or RNA analysis. Here, we describe a fluorescent assay using 4-MUG as a substrate.

1. For studying spore stages, hyphae, or infected plant tissues, grind under liquid nitrogen using a mortar and pestle (*see* **Note 8**). Then, mix ~100 μl of frozen ground tissue with an equal volume of extraction buffer (100 μl buffer for every 100 μl of packed tissue) and vortex vigorously for 30 s.
2. For hyphae or plant tissues only, lysis may alternatively be achieved by homogenization in a mechanical tissue disrupter. We have had good success by adding a fivefold volume of cold extraction buffer to ~1 mL of loosely packed hyphae in a 15-mL polypropylene tube, following by 30 s at speed 7 in a Polytron (Kinematica) with a 5-mm diameter probe.

3. Spin the lysates from steps 1 or 2 at 12000× *g* for 10 min at 4 ° C, and transfer the supernatant to a new tube.
4. Measure the protein concentrations in each extract, and then normalize all samples to the same concentration by adding extraction buffer. We normally use the Bradford or BCA methods for quantification [25].
5. For the fluorescence assay, add to a microcentrifuge 190 μL of extraction buffer containing 4-MUG and then 1 to 2 μg of the protein in 10 μL. Incubate at 37 °C for 1 h, and then add 1 mL of stop buffer and mix. For a time zero control, set up reactions in a similar manner but add the stop buffer immediately after the protein.
6. Make a standard curve by adding 200 μL of extraction buffer containing 0, 1, 4, 16, and 32 μM 4-MU to a microcentrifuge tube. Add 1 mL stop buffer and mix.
7. Measure the fluorescence at 365 mm excitation, 455 nm emission. We normally do this in a microplate reader by transferring 200 μl of each reaction into 96-well black plates.
8. Use the standard curve to calculate the specific activity of GUS enzyme (nmol MU/μg protein/min).

3.5 Promoter Mutagenesis

This is performed to identify functional regions, which are usually TF binding sites. Base changes or deletions at important locations should alter the expression pattern of the reporter gene unless redundant sites are present. As changes may be quantitative, the use of a reporter that is measured easily should be considered. These studies are often iterative, as one series of mutations may lead to testing of additional changes. These mutagenesis studies can be integrated with the gel shift experiments described in Subheading 3, **step** 7, which tests DNA for protein binding in vitro.

1. When genome data are available for related species, as is the case for *Phytophthora*, we recommend starting by aligning promoters from orthologs, i.e., phylogenetic footprinting. The logic is that functional regions should evolve more slowly than others, thus conserved regions would be candidates for mutagenesis (Fig. 2). Typically, we perform alignments of sequences from three to five species using MUSCLE or another alignment program [26]. Simultaneously, we recommend checking for over-represented motifs in co-expressed promoters using a program such as STREME [27]. The latter may be effective for cases where a TF-binding site occurs in each promoter but at unaligned sites (*see* **Note 9**).
2. Consider searching databases of known or predicted transcription factor binding sites. While this may yield false hits, the results may guide the mutagenesis strategy (*see* **Note 10**).

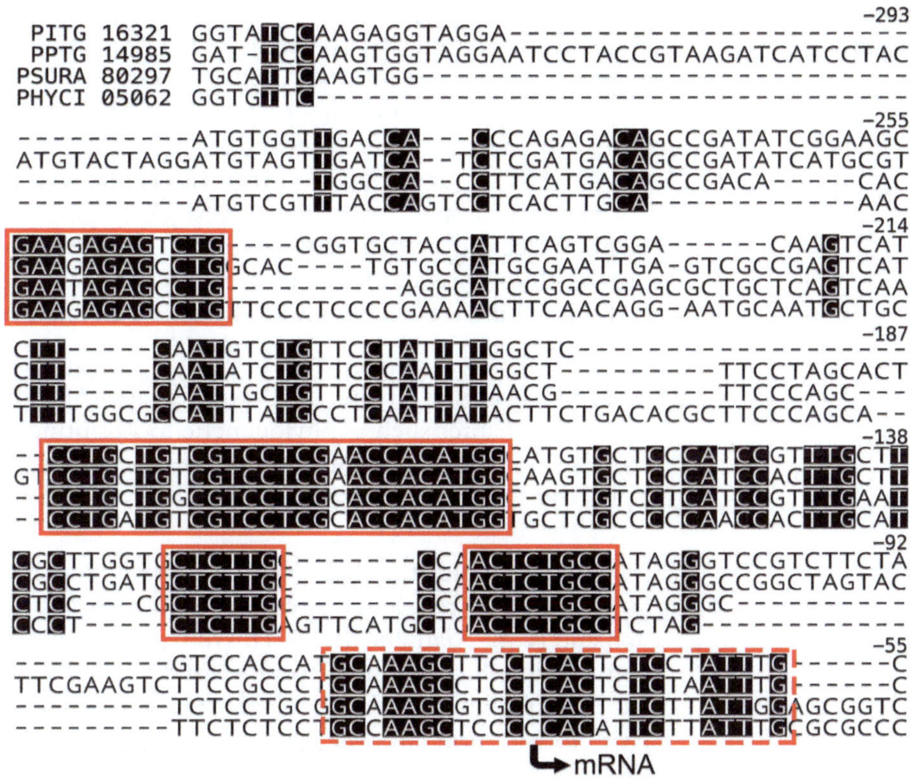

Fig. 2 Detecting conserved blocks within promoters. MUSCLE was used to align 500 nt of DNA upstream of the start codon of *P. infestans* gene PITG_16321 with its orthologs from *P. parasitica* (PPTG), *P. ramorum* (PSURA), and *P. cinnamomi* (PHYCI). The numbers correspond to the distance from the start codon in *P. infestans.* Major conserved blocks are outlined in red. These generally reside within 250 nt of the start codon and include the transcription start site (bent arrow, dashed box) and two to five upstream sites. The ends of the alignment were trimmed to save space

3. Based on results from the proceeding steps, clone modified versions into pNP-GUS. When the goal is to test regions internal to the promoter, we find it convenient to order synthetic DNA for insertion into pNP-GUS. When the goal is to generate truncate promoters at their 5′ or 3′ ends, the altered promoters are usually generated by PCR.
4. Introduce the resulting plasmids into *Phytophthora* by transformation as described in Subheading 3, **step 2**.
5. Test transformants for expression, comparing the mutagenized version with wild type, and interpret the results (*see* **Note 11**).
6. Conduct additional rounds of mutagenesis as needed to narrow down the functional sites.

3.6 Testing Motifs by Fusion to a Minimal Promoter

Many eukaryotic TF binding sites will drive transcription when cloned upstream of a minimal promoter [28]. Described below is a strategy that we have applied successfully to several motifs from *P. infestans*, using pNIFS-GUS (Fig. 3; *see* **Note 12**). The plasmid contains a minimal promoter comprised of 55-nt of DNA 5′ and 152-nt of DNA 3′ of the major transcription start site of the *PiNIFS* gene (Genbank accession AY751569), plus the *nptII* marker for selecting transformants. GUS expression does not occur in the absence of additional upstream sequences, but usually results if a site binding a TF with an activation domain has been added.

1. Select the motif for testing. This might be identified by methods in this chapter such as phylogenetic footprinting, deletion analysis, or EMSA. Motifs might also be identified from approaches such as over-representation analysis or SELEX-seq [15, 29].
2. Design a double-stranded oligonucleotide containing the motif, including restriction sites compatible with the vector at each end (e.g., *Xba*I and *Bam*HI). We typically design a 35-nt fragment in which the motif of interest resides close to the 5′ end. The oligonucleotides should be staggered such that cohesive ends will form after the following annealing step.

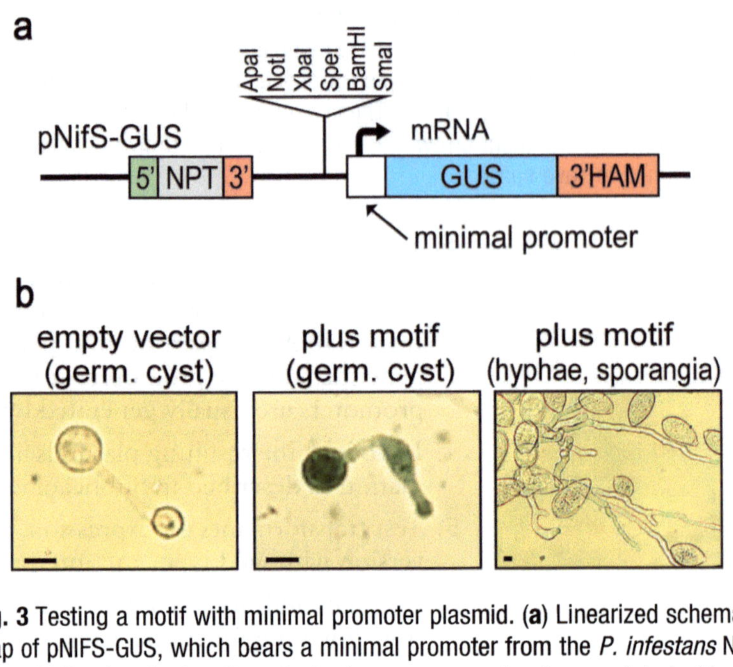

Fig. 3 Testing a motif with minimal promoter plasmid. (**a**) Linearized schematic map of pNIFS-GUS, which bears a minimal promoter from the *P. infestans* NIFS gene. (**b**) Testing the function of a 6-nt sequence previously associated with cyst germination based on motif over-representation analysis of co-expressed genes. When cloned into the *Xba*I and *Sma*I sites of the polylinker, GUS was detected in germinating cysts but not in sporangia or vegetative hyphae. Bars equal 5 μm

3. After obtaining the oligonucleotides, resuspend each in TE to a concentration of 100 μM.
4. Combine 10 μL of each strand in a single tube, and heat at 95 °C for 2 min. The mixture is then cooled to 4 °C over about 15 min. This can be done in a thermal cycler with a ramp program, or in steps by incubating the tube at 72 °C for 4 min, 37 °C for 4 min, 25 °C for 4 min, and then holding the tube at 4 °C.
5. Following the steps in Sect. 3.1.4, digest pNIFS-GUS with the appropriate restriction enzymes, combine with the annealed oligonucleotides, ligate, transform *E. coli,* and identify colonies containing the insert.
6. Transform the resulting plasmid to *Phytophthora* by transformation as described in Subheading 3, **step 2**.
7. Observe the pattern of GUS expression in the transformants and interpret the results (*see* **Note 13**).

3.7 Gel Shift Assays (EMSA)

This helps link TFs to their targets in the promoter, based on the principle that a protein-nucleic acid complex has less electrophoretic mobility compared to the free nucleic acid. The assay is straightforward although some TFs may exhibit weak binding and be challenging to study (*see* **Note 14**).

1. Identify the DNA to be tested. When testing a single motif, we normally use double-stranded DNA targets (dsDNAs) having a minimum size of about 35-nt, in which the site of interest (TF binding sites are typically 6–12 nt) is flanked by additional bases. These dsDNAs are typically made by annealing a Cy5-labeled oligonucleotide with a complementary unlabeled sequence. If greater sensitivity is desired, both strands may be labeled. Other fluorescent tags such as Cy3 or fluorescein, or a 5' ^{32}P label added using T4 kinase, may also be used. A dsDNA fragment lacking the target sequence should also be designed as a control, to be used as a nonspecific competitor in the binding reaction (*see* **Note 15**).
2. Obtain your protein sample, either a purified TF or a nuclear extract. For the latter, harvest and resuspend tissue in Nuclear Buffer 1. Homogenize on ice using 0.5-mm glass beads in a Bead-Beater, employing three strokes of 30 s, each separated by a 2 min cooling period. Subsequent steps should be at 4 °C. Add an equal volume of Nuclear Buffer 1 containing 1% Triton X-100 to the lysed tissues and incubate for 10 min. Pellet the nuclei at 8000 × *g* for 5 min and then wash twice with 25 mL Nuclear Buffer 1. Resuspend the pellet in 1.5 times the pellet volume of Nuclear Buffer 2 containing 0.5 mM PMSF. Next, add 4.75 volumes of water, followed by 1.25 volumes of 3 M KCl added dropwise. Stir gently for 30 min. Centrifuge for

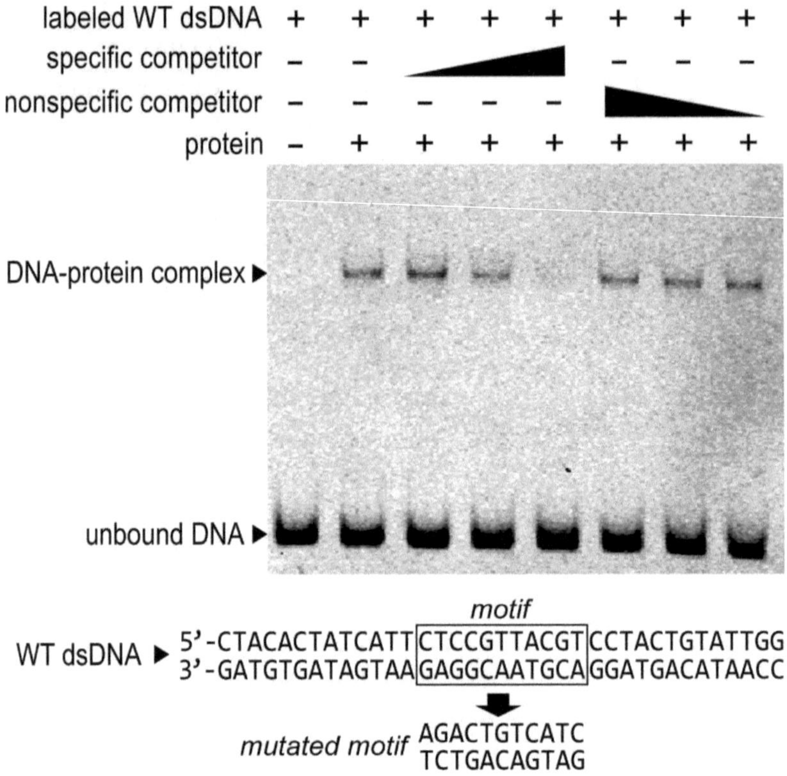

Fig. 4 Fluorescent EMSA. A TF was incubated with a 37-nt double-stranded Cy5-labeled oligonucleotide containing a candidate motif and subjected to electrophoresis. As indicated above the gel, some lanes included unlabeled 37-mer as a specific competitor or an unlabeled nonspecific competitor, which was the 37-mer with the motif mutated. The triangles represent competitors added at 1, 10, and 100-fold the concentration of the Cy5-labeled target. The sequences of the unmutated (WT) and mutated 37-mers are represented beneath the gel. Specific binding of the TF to the motif is indicated since only the specific competitor diminished the intensity of the retarded band, which represents the DNA-protein complex

30 min at 16,000× *g* to collect the supernatant, which should be dialyzed overnight against Nuclear Buffer 3. Finally, clarify by centrifugation at 16,000× *g* for 30 min. Store at −80 °C until ready to use.

3. Resuspend each oligonucleotide at 100 μM in TE buffer.

4. In separate tubes, anneal the three pairs of oligonucleotides. The first will be the Cy5-labeled target and its unlabeled complement; the second will be a specific competitor made from the same two sequences but without Cy5; and the third will be a nonspecific competitor control without Cy5. The latter could be random DNA of the same size, or the same sequence used for the Cy5 oligo but with the predicted target site altered, as

shown in Fig. 4. Set up the annealing reaction by combining 1 μL of each oligonucleotide, 1 μL of 1 M Tris pH 7.9, 2.5 μL of 50 mM KCl, 1 μL of 100 mM $MgCl_2$, and 43.5 μL of dH_2O. After heating for 3 min in a heat block at 95 °C, cool slowly by placing the block at room temperature (on the benchtop) for 1 h. Alternatively, the mixture can be cooled in steps in a thermal cycler or a series of heat blocks as in Sect. 3.6.4. Run a small amount of the reaction on a 5% polyacrylamide gel to confirm the generation of a band of the expected size.

5. Cast the 5% polyacrylamide gel. We normally use 7-cm tall gels with 1.5-mm spacers and wells 5-mm in width.
6. Prepare the DNA and protein samples, keeping each at 4 °C. The optimal concentrations of Cy5-labeled DNA and protein will vary with each transcription factor, so we advise that these be measured through titration experiments (*see* **Note 16**). A typical starting point would be a final concentration of 15 nM DNA and 150 nM purified protein (or 6 μg of nuclear extract protein) in a 20 μl reaction. When a nuclear extract is used, we include 1.3 μg of poly (dI-dC) to reduce non-specific binding.

   Once the optimal concentrations are established, the assays are assembled by adding the appropriate DNA to 10 μl of Binding Buffer A, the protein to 10 μl of Binding Buffer B, and then combining the two tubes. This should result in the following mixtures:

   (a) Cy5-labeled dsDNA alone.
   (b) Cy5-labeled dsDNA plus protein.
   (c) Cy5-labeled dsDNA with an equal amount of the same unlabeled DNA, plus protein.
   (d) Cy5-labeled dsDNA with a tenfold excess of the same unlabeled DNA, plus protein.
   (e) Cy5-labeled dsDNA with a 100-fold excess of the same unlabeled DNA, plus protein.
   (f) Cy5-labeled dsDNA with an equal amount of control DNA, plus protein.
   (g) Cy5-labeled dsDNA with a tenfold excess of control DNA, plus protein.
   (h) Cy5-labeled dsDNA with a 100-fold excess of control DNA, plus protein.

   The concentration of the Cy5-labeled dsDNA should remain the same in each of the eight reactions. Reactions c to e represent those with the specific competitor, while f to h represent those with the nonspecific competitor.

7. Incubate at room temperature for 15 min in the dark.
8. Add 2 μL of Orange Loading Dye to each reaction and load 19 μL on the gel. Run for 30 min at 15 V/cm (110 V for a 7-cm gel) at room temperature (*see* **Note 17**).
9. Detect the signal using a scanner. For Cy5, we use 633 nm excitation, 670 nm emission.
10. Interpret the results. Binding will be signaled by a retarded band (Fig. 4). Specific binding will be indicated if the signal drops off proportionately with the amount of unlabeled specific competitor. The nonspecific competitor should have less of an effect, although it might partially reduce the intensity of the retarded band since most TFs have some nonspecific DNA-binding ability. Depending on the application, the results may lead to an additional round of EMSA with altered dsDNAs to refine the definition of the TF binding site.

4 Notes

1. GUS should provide a stronger result than a fluorescent reporter since the former benefits from enzymatic amplification of its signal. Histochemical staining for GUS also does not require the instrumentation required for measuring fluorescence. However, the staining profiles may be deceptive since the assay is only semiquantitative. Also, weak signals may be detected with difficulty in small cells such as zoospores. One reason for this is that the initial reaction product, 5-bromo-4-chloro-indoxyl, must dimerize with its tautomer in an oxidation reaction that forms the blue precipitate, 5,5′-dibromo-4,4′-dichloro-indigo. The initial product may diffuse from the cell prior to dimerization if GUS levels are very low, although the ferro- and ferricyanates are included to accelerate dimerization by catalyzing the oxidation reaction [30]. Do not be tempted to leave these out of the reaction! This problem does not occur with fluorescent protein reporters, which unlike GUS are amenable to live imaging. One issue with both types of reporters is that they have long half-lives, with GUS appearing to be more stable than GFP at least in plants [31]. This helps to augment their signals but may complicate some temporal studies. Destabilized reporters have been developed for some taxa but not yet for oomycetes [32].
2. Most promoters of protein-coding genes in eukaryotes range in size from a few hundred bases to about 1 kb [33]. This does not include distal enhancer sites which in mammals may be tens of kb 5′ or 3′ of the coding sequences [34]; whether such sites exist in oomycetes is unknown. We generally start our

experiments by testing the 500 bases upstream of the translation start site or the intergenic region, whichever is smaller. Phylogenetic footprinting may also suggest the size of the functional promoter (Sec. 3.5.1), which can be tested by 5′ deletion analysis. Identifying the transcription start site is also recommended. While this can be defined by traditional techniques such as 5' RACE, it can be estimated by mapping RNA-seq reads, which are now widely available for many *Phytophthora* spp. including *P. infestans* [35].

3. These times may vary depending on the strain or species. While *P. infestans* normally sporulates as cultures age, other species may require treatments with light or starvation. The goal is to obtain young synchronous sporangia that will effectively release zoospores within a fairly narrow window of time (typically 1.5 to 2 h after chilling for *P. infestans*). A simple 10 °C chamber for chilling can be made by filling a plastic tub with ice, covering the ice with a 3 to 6-mm acrylic sheet upon which the plates of sporangia are positioned, and then covering the apparatus with an inverted tub.
4. These parameters work well for most strains of *P. infestans.* Changes in the electrical conditions may be needed for species making larger or smaller zoospores to achieve the optimal time constant and voltage drop across the cell.
5. To prevent contamination, we often add an antibacterial (e.g., 50 μg/mL penicillin G) and antifungal (e.g., 40 units/mL nystatin) to the media.
6. This concentration works well for most strains of *P. infestans.* Adjustments may be needed depending on the species, strain, or batch of chemical.
7. It is prudent to examine multiple transformants to avoid artifacts due to position effects or variation in copy number [36]. This will also reduce the possibility of drawing a false conclusion if the reporter has integrated adjacent to another promoter. Alternatively, use CRISPR [37] to insert the reporter into a safe harbor [38]. Promoter swaps using CRISPR have been reported in *Phytophthora sojae,* but note that these run the risk of being lethal if the targeted gene was essential [39].
8. Protein extracted from ~150 μg of tissue is usually sufficient. For hyphae, it is convenient to use either liquid cultures or grow the strain on top of a 0.4 μm pore size polycarbonate filter laid on top of agar media; the hyphae can be easily peeled off the membrane.
9. It is essential to use orthologs and not paralogs, since many genes belong to families in which members may show diverse patterns of transcription. Moreover, the promoters should be

suitably distant (typically <70% identical in alignments) to allow non-conserved and thus likely nonfunctional sites to be distinguished from conserved blocks. We usually start by aligning promoters from at least five species, and then eliminate species if needed to optimize the discrimination of the analysis. Alignments are usually effective at revealing conserved blocks as shown in Fig. 2, although modifying gap penalties may help to optimize the alignment. However, DNA rearrangements may have moved TF binding sites to locations that would remain unaligned. Such rearrangements are not uncommon, even within a species [40]. Programs such as STREME, which uses expectation maximization to identify over-represented words, may succeed in identifying conserved motifs even within a set of rearranged promoters [27]. This approach may also be useful for studying promoters where orthologs are unavailable, as is the case for some effectors. In such cases, over-represented motifs (putative TF binding sites) may be detected within promoters showing similar patterns of activity [15].

10. Several papers have reported predicted or experimentally proven regulatory sites within *Phytophthora* promoters [15, 41–43]. There may also be utility in searching the extensive databases that exist for TFs from model plants, animals, and fungi [44]. Our experimental analysis of the DNA binding sites of over 100 *P. infestans* TFs indicates that many resemble those of TFs in the same family from other kingdoms. However, be aware that many hits in databases will be false positives [45].

11. Possible results include a loss of stage or condition-specific transcription due to altering the binding of a stage-specific activator, expression in a new pattern due to eliminating a repressor site, or a quantitative reduction if a motif bound by a general transcription factor was deleted. No change means either that the mutated site was nonfunctional or redundant.

12. This was named p74NIF in our prior publications based on expressed sequence tag data that indicated that it included 74-nt of DNA 5′ of the transcription site [14]. More accurate techniques have shown that the start site is 27–29 nt further upstream.

13. There are limitations to this method. While sites that bind TFs with activation domains usually yield visible GUS expression, the signal is usually lower than that obtained with the complete promoter due to the absence of binding sites for general transcription factors. It is also possible that no reporter may be made if the TF binding the motif must operate in combination with other regulatory elements. In addition, TFs acting as repressors will not yield results in this assay.

14. Our protocol has succeeded with multiple TFs, but no single combination of protein extraction, binding, or electrophoresis conditions may work well for all [46]. For example, protein-nucleic acid interactions are sensitive to pH and the levels of mono- and divalent salts. While our protocol uses a Tris-HEPES buffer at an approximately physiological ionic strength and pH, and on ice, some TFs may bind better under alternative conditions. Our inclusion of bases flanking the putative TF binding site is meant to strengthen the DNA-protein interaction as the TF might "sit down" better on a longer target. However, the longer template may bear non-specific binding sites, which can be a drawback when binding the DNA to a complex nuclear extract. The longer targets also migrate more slowly on gels which provides more time for the desired complex to dissociate, although this could be mitigated by running the gel in the cold or for a shorter time. Another example of a challenge is when binding occurs only when the TF includes a post-translational modification such as phosphorylation, and thus it may be prudent to include a phosphatase inhibitor during extractions.
15. Poly (dI-dC) can also be used as a nonspecific competitor [47].
16. Titration can improve the results of EMSA, particularly when using a purified TF [46]. If the protein concentration is too low, a signal will not be detected while if too high, the DNA competitor may not have an effect. We typically perform titrations using DNA at about 25 nM in Binding Buffer A and proteins ranging from 0 to 300 nM in Binding Buffer B.
17. It may be useful to pre-run the gel until the current stabilizes (20–30 min) to remove persulfate ions which may weaken the protein-DNA complex. If a certain ion or other molecule was found to be important for binding, this could be included in the electrophoresis buffer to stabilize the complex. This is why glycerol is present both in the running buffer and gel, as this fortifies most protein-DNA interactions.

Acknowledgements

This work was supported by grants from the US National Science Foundation and the USDA National Institute of Food and Agriculture to HSJ.

References

1. Erwin DC, Ribeiro OK (1996) *Phytophthora* diseases worldwide. APS Press, St. Paul, Minn
2. Gamboa-Melendez H, Huerta AI, Judelson HS (2013) bZIP transcription factors in the oomycete *Phytophthora infestans* with novel

DNA-binding domains are involved in defense against oxidative stress. Eukaryot Cell 12: 1403–1412

3. Xiang Q, Judelson HS (2014) Myb transcription factors and light regulate sporulation in the oomycete *Phytophthora infestans*. PLoS One 9:e92086
4. Leesutthiphonchai W, Judelson HS (2018) A MADS-box transcription factor regulates a central step in sporulation of the oomycete *Phytophthora infestans*. Mol Microbiol 110: 562–575
5. Sheng YT, Wang YL, Meijer HJG, Yang XY, Hua CL, Ye WW et al (2015) The heat shock transcription factor PsHSF1 of *Phytophthora sojae* is required for oxidative stress tolerance and detoxifying the plant oxidative burst. Environ Microbiol 17:1351–1364
6. Zhang M, Lu J, Tao K, Ye WW, Li AN, Liu XY et al (2012) A Myb transcription factor of *Phytophthora sojae*, regulated by MAP kinase PsSAK1, is required for zoospore development. PLoS One 7:e40246
7. Sheng YT, Lin L, Chen H, Pu THZ, Liu XY, Dong SM et al (2021) The bZIP transcription factor PsBZP32 is involved in cyst germination, oxidative stress response, and pathogenicity of *Phytophthora sojae*. Phytopathol Res 3:1
8. Blanco FA, Judelson HS (2005) A bZIP transcription factor from *Phytophthora* interacts with a protein kinase and is required for zoospore motility and plant infection. Mol Microbiol 56:638–648
9. Bottin A, Larche L, Villalba F, Gaulin E, Esquerre-Tugaye M-T, Rickauer M (1999) Green fluorescent protein (GFP) as gene expression reporter and vital marker for studying development and microbe-plant interaction in the tobacco pathogen *Phytophthora parasitica* var. *nicotianae*. FEMS Microbiol Lett 176:51–56
10. Ah Fong AM, Judelson HS (2003) Cell cycle regulator Cdc14 is expressed during sporulation but not hyphal growth in the fungus-like oomycete *Phytophthora infestans*. Mol Microbiol 50:487–494
11. Lee MH, Chiu CM, Roubtsova T, Chou CM, Bostock RM (2010) Overexpression of a redox-regulated cutinase gene, *MfCUT1*, increases virulence of the brown rot pathogen *Monilinia fructicola* on *Prunus* spp. Mol Plant Microbe In 23:176–186
12. Tani S, Judelson HS (2006) Activation of zoosporogenesis-specific genes in *Phytophthora infestans* involves a 7-nucleotide promoter motif and cold-induced membrane rigidity. Eukaryot Cell 5:745–752
13. Sornkom W, Asano K, Sone T (2015) Use of native promoter-eGFP as a gene reporter on onion epidermis to analyze gene expression of AVR-pia, an avirulence effector of rice blast pathogen. Eng J 19:85–93
14. Ah Fong A, Xiang Q, Judelson HS (2007) Architecture of the sporulation-specific Cdc14 promoter from the oomycete *Phytophthora infestans*. Eukaryot Cell 6:2222–2230
15. Roy S, Kagda M, Judelson HS (2013) Genome-wide prediction and functional validation of promoter motifs regulating gene expression in spore and infection stages of *Phytophthora infestans*. PLoS Pathog 9:e1003182
16. Steiner S, Pfannschmidt T (2009) Fluorescence-based electrophoretic mobility shift assay in the analysis of DNA-binding proteins. Methods Mol Biol 479:273–289
17. Heffler MA, Walters RD, Kugel JF (2012) Using electrophoretic mobility shift assays to measure equilibrium dissociation constants: GAL4-p53 binding DNA as a model system. Biochem Mol Biol Edu 40:383–387
18. Jolma A, Yan J, Whitington T, Toivonen J, Nitta KR, Rastas P et al (2013) DNA-binding specificities of human transcription factors. Cell 152:327–339
19. Andersson R, Sandelin A (2020) Determinants of enhancer and promoter activities of regulatory elements. Nat Rev Genet 21:71–87
20. Jones DM, Vandepoele K (2020) Identification and evolution of gene regulatory networks: insights from comparative studies in plants. Curr Opin Plant Biol 54:42–48
21. Zambelli F, Pesole G, Pavesi G (2013) Motif discovery and transcription factor binding sites before and after the next-generation sequencing era. Brief Bioinform 14:225–237
22. Caten CE, Jinks JL (1968) Spontaneous variability in isolates of *Phytophthora infestans*. I Cultural variation Can J Bot 46:329–348
23. Ah-Fong AM, Judelson HS (2011) Vectors for fluorescent protein tagging in *Phytophthora*: tools for functional genomics and cell biology. Fungal Biol 115:882–890
24. Ah-Fong AMV, Kagda M, Judelson HS (2018) Illuminating *Phytophthora* biology with fluorescent protein tags. Plant pathogenic fungi and oomycetes: methods and protocols. In: Methods in Molecular Biology, vol 1848. Humana Press Inc, Totowa, NJ, pp 119–129
25. Noble JE, Bailey MJA (2009) Quantitation of protein. Method Enzymol 463:73–95
26. Edgar RC (2004) MUSCLE: multiple sequence alignment with high accuracy and high throughput. Nucleic Acids Res 32:1792–1797

27. Bailey TL (2021) STREME: accurate and versatile sequence motif discovery. Bioinformatics 37:2834–2840
28. Smith RP, Riesenfeld SJ, Holloway AK, Li Q, Murphy KK, Feliciano NM et al (2013) A compact, in vivo screen of all 6-mers reveals drivers of tissue-specific expression and guides synthetic regulatory element design. Genome Biol 14:R72
29. Smaczniak C, Angenent GC, Kaufmann K (2017) SELEX-Seq: a method to determine DNA binding specificities of plant transcription factors. Plant Gene Regulatory Networks: Methods Mol Biol 1629:67–82
30. Lojda Z (1970) Indigogenic methods for glycosidase. 1. An improved method for beta-d-glucosidase and its application to localization studies of intestinal and renal enzymes. Histochemie 22:347–361
31. de Ruijter NCA, Verhees J, van Leeuwen W, van der Krol AR (2003) Evaluation and comparison of the GUS, LUC and GFP reporter system for gene expression studies in plants. Plant Biol 5:103–115
32. Li XQ, Zhao XN, Fang Y, Jiang X, Duong T, Fan C et al (1998) Generation of destabilized green fluorescent protein transcription reporter. J Biol Chem 273:34970–34975
33. Le NQK, Yapp EKY, Nagasundaram N, Yeh HY (2019) Classifying promoters by interpreting the hidden information of DNA sequences via deep learning and combination of continuous FastText N-grams. Front Bioeng Biotech 7:305
34. Panigrahi A, O'Malley BW (2021) Mechanisms of enhancer action: the known and the unknown. Genome Biol 22:108
35. Ah-Fong AM, Kim KS, Judelson HS (2017) RNA-seq of life stages of the oomycete *Phytophthora infestans* reveals dynamic changes in metabolic, signal transduction, and pathogenesis genes and a major role for calcium signaling in development. BMC Genomics 18:198
36. Judelson HS, Dudler R, Pieterse CMJ, Unkles SE, Michelmore RW (1993) Expression and antisense inhibition of transgenes in *Phytophthora infestans* is modulated by choice of promoter and position effects. Gene 133:63–69
37. Ah-Fong A, Boyd AM, Matson MEH, Judelson HS (2021) A Cas12a-based gene editing system for *Phytophthora infestans* reveals monoallelic expression of an elicitor. Molec Plant Pathol 22:737–752
38. Kelly JJ, Saee-Marand M, Nystrom NN, Evans MM, Chen YX, Martinez FM et al (2021) Safe harbor-targeted CRISPR-Cas9 homology-independent targeted integration for multimodality reporter gene-based cell tracking. Sci Adv 7:eabc3791
39. Ochola S, Huang J, Ali H, Shu HD, Shen DY, Qiu M et al (2020) Editing of an effector gene promoter sequence impacts plant-*Phytophthora* interaction. J Integr Plant Biol 62:378–392
40. Matson MEH, Liang Q, Lonardi S, Judelson HS (2022) Karyotype variation, spontaneous genome rearrangements affecting chemical insensitivity, and expression level polymorphisms in the plant pathogen *Phytophthora infestans* revealed using its first chromosome-scale assembly. PLoS Pathog 18:e1010869
41. Roy S, Poidevin L, Jiang T, Judelson HS (2013) Novel core promoter elements in the oomycete pathogen *Phytophthora infestans* and their influence on expression detected by genome-wide analysis. BMC Genomics 14:106
42. Xiang Q, Kim KS, Roy S, Judelson HS (2009) A motif within a complex promoter from the oomycete *Phytophthora infestans* determines transcription during an intermediate stage of sporulation. Fungal Genet Biol 46:400–409
43. Lin L, Ye WW, Wu JW, Xuan MR, Li YF, Gao J et al (2018) The MADS-box transcription factor PsMAD1 is involved in zoosporogenesis and pathogenesis of *Phytophthora sojae*. Front Microbiol 9:2259
44. Castro-Mondragon JA, Riudavets-Puig R, Rauluseviciute I, Lemma RB, Turchi L, Blanc-Mathieu R et al (2022) JASPAR 2022: the 9th release of the open-access database of transcription factor binding profiles. Nucleic Acids Res 50:D165–DD73
45. Wasserman WW, Sandelin A (2004) Applied bioinformatics for the identification of regulatory elements. Nat Rev Genet 5:276–287
46. Hellman LM, Fried MG (2007) Electrophoretic mobility shift assay (EMSA) for detecting protein-nucleic acid interactions. Nat Protoc 2: 1849–1861
47. Larouche K, Bergeron MJ, Leclerc S, Guerin SL (1996) Optimization of competitor poly (dI-dC) poly(dI-dC) levels is advised in DNA-protein interaction studies involving enriched nuclear proteins. BioTechniques 20: 439–444

Chapter 2

RNA Silencing Strategies in *Phytophthora*: Experimental Guidelines and Insights

Stephen C. Whisson, Lydia R. J. Welsh, and Ramesh R. Vetukuri

Abstract

RNA silencing is a core cellular process that acts to defend the genome against potentially damaging genetic elements such as viruses and transposons. It has been extensively characterized in many eukaryotes and exploited as a tool for determining gene function through removing the activity of specific genes. It has also been used in *Phytophthora* species to reveal genes involved in different lifecycle stages. In this chapter, we provide guidelines and outline considerations for carrying out RNA silencing experiments in *Phytophthora*.

Key words *Phytophthora*, RNA silencing, RNA interference, sRNA, Gene expression, Heterochromatin

1 Introduction

Determining the role(s) of specific genes and proteins in the biology of an organism is a goal of many research projects. Where they are available, chemical inhibitors can be used to block the activity of a specific protein, or target genes can be overexpressed in transgenic lines of the organism of interest. However, it is also very informative to remove the activity of a gene to determine the importance of its encoded protein or its transcribed RNA. This can be achieved through targeted disruption using strategies such as CRISPR/Cas. Alternatively, genes can be silenced using RNA silencing.

RNA silencing, often called RNA interference, is a cellular mechanism that is broadly conserved in eukaryotes. Its biological function is to defend genomes against viruses, transposable elements, and other aberrant RNAs [1]. Three proteins are at the core of RNA silencing: Dicer (or Dicer-like; DCL), Argonaute (AGO), and RNA-dependent RNA polymerase (RDRP). Dicer has two RNaseIII domains and acts to digest double-stranded RNAs (dsRNA) into short interfering RNAs (often called small RNAs; sRNAs) [2]. These are unwound and bound by the

Monica L. Gerth and Rosie E. Bradshaw (eds.), *Phytophthora: Methods and Protocols*, Methods in Molecular Biology, vol. 2892, https://doi.org/10.1007/978-1-0716-4330-3_2,

RNA-induced silencing complex that contains the AGO protein. The AGO-sRNA complex then targets complementary messenger RNA (mRNA) and degrades it [3]. RDRP can act on degraded mRNA fragments to drive the formation of more dsRNA and, thus, more sRNAs [4, 5]. This process acts at the post-transcriptional level, that is, after mRNA has been transcribed. However, RNA silencing can also link to chromatin modifications such as cytosine or histone methylation, leading to transcriptional silencing [6, 7].

In *Phytophthora*, silencing of introduced transgenes has been known since the early experiments to transform *P. infestans* [8]. However, the underlying basis of silencing was not understood for several years [9]. The availability of sequenced genomes for many *Phytophthora* species has enabled the identification of the core components for RNA silencing in *Phytophthora* and other oomycetes [10–15]. The *Phytophthora* species analyzed to date possess at least one DCL protein, multiple AGO proteins, and one RDRP protein. Sequencing and northern blot analysis of sRNAs from *Phytophthora* species has revealed the presence of sRNAs ranging in size from 20 to 35 nt, but with a marked enrichment for sRNAs of 21 nt, 25 nt and 26 nt [10–12, 14, 16, 17]. *Phytophthora* sRNAs map predominantly to repetitive sequences and transposons in *Phytophthora* genomes, but can also target duplicated genes such as effectors [10–12, 14, 17].

RNA silencing in *Phytophthora* has been exploited to determine the action of many genes involved in different stages of the lifecycle, such as sporulation, appressorium formation, and pathogenesis, with *P. infestans* and *P. sojae* being the two species where silencing has been used most often [18–22]. The initial studies that used RNA silencing in *Phytophthora* relied on stable transformation, which can be technically challenging and of low efficiency. However, other RNA silencing strategies have since been developed that can circumvent the technical challenges associated with stable transformation of *Phytophthora*. For example, treatment of protoplasts with in vitro synthesized dsRNA demonstrated that dsRNA could be taken up and silence genes in *P. infestans*, with the transient silencing effect persisting up to 14 days [23]. DsRNAs homologous to *Phytophthora* genes can be expressed in host plants, and (presumably) sRNAs can be transferred to invading *Phytophthora* hyphae to silence the target gene in the pathogen (called host-induced gene silencing; HIGS) [24–27]. The precise mechanism of plant to *Phytophthora* (trans-boundary) dsRNA or sRNA transfer is yet to be determined. More recently, it has been demonstrated that spraying plant leaves with in vitro synthesized dsRNA homologous to *Phytophthora* genes can silence genes in the pathogen (called spray-induced gene silencing; SIGS) [24, 28–30]. This development not only opens up a new avenue for disease control but enables the determination of gene function in pathogenicity.

Details of how to carry out assays such as SIGS and dsRNA treatment of protoplasts, *Phytophthora* transformation, and downstream methods such as reverse transcribed quantitative PCR are detailed elsewhere and in this volume [23, 31–34]. As such, this chapter will provide guidelines and considerations for carrying out RNA silencing experiments with *Phytophthora*, with a focus on silencing in stable transformed lines.

2 Advantages and Disadvantages of RNA Silencing for Determining Gene Function in *Phytophthora*

Each silencing strategy can yield different levels of target transcript knock-down in *Phytophthora*. HIGS, SIGS, and protoplast dsRNA treatment typically lead to partial and transient silencing of the target gene [23, 28, 35]. An advantage of this is that these strategies can be used for silencing genes that are otherwise lethal when fully inactivated. Both SIGS and protoplast dsRNA treatment can also be adapted to higher throughput screening of genes for phenotypes from RNA silencing. The main disadvantages to these strategies are that the target gene does not remain silenced beyond the assay time, signifying that new dsRNA treatments must be established for each experiment. Secondly, as stated earlier, these strategies typically lead to partial silencing of the target gene. That is, the target mRNA levels are reduced to a fraction of the untreated controls, which may or may not be sufficient to reveal an informative phenotype.

By comparison, transgene-mediated silencing often results in near-complete transcriptional inactivation of the target gene, effectively yielding a null mutant. This can make phenotypic characterization of silenced lines more straightforward than using partially silenced lines. Silencing of essential genes can potentially also be achieved with this strategy, although only partial silencing may be achieved. A further advantage is that the numbers of transformed lines required to identify those that are successfully silenced are typically lower than that for biallelic gene knockout using CRISPR/Cas, meaning that high-efficiency transformation is not as critical. For example, silencing of a lytic polysaccharide monooxygenase in *P. infestans* revealed that approximately 30% of transformed lines were silenced for the target gene [21]. In comparison, the most recent developments in the CRISPR-Cas12a system in *P. infestans* revealed that typically 5% of transformed lines will show biallelic editing at the target locus [36].

Although transgene-mediated silencing does yield stronger levels of transcript knock-down, it is typically not as high-throughput as SIGS or dsRNA treatment of protoplasts. Although this silencing strategy is often referred to as being stable, the silencing of the target gene can change over time [37].

3 Plasmid Vectors for Transgene-Mediated Silencing

Transgene-mediated silencing in *Phytophthora* has been shown to be initiated by overexpression of sense or antisense copies of target genes. However, these approaches can be inefficient for generating silenced lines [37]. Most transgene-mediated silencing experiments tend to use inverted repeats (hairpin), with a sense and antisense copy of the target sequence separated by a short spacer, typically an intron. When transcribed, the resulting mRNA folds back on itself to form dsRNA, which can then be cleaved into sRNAs by dicer, thus initiating silencing. An example of this is the pSTORA plasmid, in which expression of the inverted repeat is driven by the strong constitutive *Ham34* promoter from *Bremia lactucae* [38] and the sense and antisense copies of the target sequence are separated by a 71 bp intron from the *P. infestans Pic20* gene [37, 39]. This plasmid vector also contains the neomycin phosphotransferase (*nptII*) gene for G418 (geneticin) resistance to enable selection of transformed *Phytophthora* colonies.

An alternative to using an inverted repeat construct to initiate silencing has been to exploit the transposon complement of a host genome. For example, silencing of an endogenous effector gene was achieved using a transcriptional fusion of a short interspersed element (SINE) from *P. infestans* with the *Avr3a* effector gene [40]. Presumably, silencing of the SINE by endogenous sRNAs also silenced the fused and endogenous copies of *Avr3a*. In that study, 3′ fusion of the SINE to *Avr3a* (*Avr3a*-SINE) yielded the best silencing results. A disadvantage of this silencing strategy is that it will likely be limited to *P. infestans*, the source species for the SINE.

Although silencing is mediated by sRNAs, the use of sRNAs as the molecules to initiate silencing has not always been as effective as longer dsRNA sequences in *Phytophthora*. A short hairpin RNA of 21 nt was used to initiate silencing of the NIFC1 gene but was found to be approximately 50-fold less effective than a longer inverted repeat [41]. However, sRNAs were effective at silencing genes in *P. capsici* when applied in SIGS experiments [24]. Thus, dsRNA sequences of 200 bp to 1 kb in size are typically used to initiate silencing in *Phytophthora*. This can be the entire length of many *Phytophthora* genes, especially effector coding genes.

4 Transformation Method

There are multiple methods for transforming *Phytophthora*, such as *Agrobacterium* transformation, microprojectile bombardment, electroporation of zoospores, and polyethylene glycol (PEG)/ $CaCl_2$ transformation of protoplasts. Each transformation method results in different patterns of transgene insertion and copy number. A side-by-side comparison of zoospore electroporation, microprojectile bombardment, and PEG/$CaCl_2$-protoplast transformation methods in *P. infestans* showed that the PEG/-$CaCl_2$-protoplast method yielded the highest proportion of silenced transformants when combined with an inverted repeat plasmid construct [37]. The same study revealed that transgene copy number and silencing frequency were correlated. *Phytophthora* transformation methods are well documented elsewhere, and also in this volume [34].

5 Assessment of Gene Expression

Many silencing experiments involve determining the role of a gene that exhibits differential expression at a specific stage of the *Phytophthora* lifecycle. From transcriptome experiments, it is clear that even genes that exhibit strong lifecycle stage-specific expression also have a basal level of expression at other stages [42–44]. When assessing levels of silencing, to most accurately determine the level of silencing, the stage or conditions where the target gene would normally be most highly expressed should be used to determine if the gene is silenced. This can be particularly important when identifying partial silencing of a target gene, where the silencing may not be apparent if the most appropriate lifecycle stage is not assessed. For example, if a target gene is most highly expressed in sporangia, then conditions under which sporangia are normally formed should be assessed for silencing of the target gene.

The level of target gene expression should be assessed in all lines used in a silencing experiment to ensure that the level of silencing can be correlated with any observed phenotypes. Reverse transcription quantitative PCR (RT-qPCR) is an ideal method for this, as it can yield information as to what level the target gene has been silenced, when measured against the expression of constitutively expressed internal control genes. Internal control genes that have been used for RT-qPCR in *Phytophthora* are actin A, ribosomal protein S3A, a RAS GTPase, casein kinase, and nonsense-mediated mRNA decay protein 3 [13, 21, 45]. These control genes have typically been selected from transcriptome profiling experiments as having no or minimal change in transcript level in all lifecycle stages analyzed. Candidate silenced lines from transformation with inverted repeat constructs can often have no detectable transcript from the target gene, although partially silenced lines are also often recovered [37]. To accurately determine the level of target gene expression, the primers used for RT-qPCR should be located outside the gene region used to design the dsRNA. This can be challenging when using the entire gene sequence to produce the dsRNA to initiate silencing. In this case, a forward primer located near the 3′ end of the gene, combined with a reverse primer in the 3′ untranslated region of the target gene, can be used to determine the mRNA level from the target gene. Additionally, a similar strategy can be used to determine if the inverted repeat construct is being expressed while the target gene is silenced [46]. This situation can occur in *Phytophthora* and can confuse analyses of target gene expression if RT-qPCR primers are not placed outside the dsRNA-targeted region. Since silencing levels can vary from one experiment to another, even with transgene-mediated "stable" silenced lines, each time transformants are phenotyped, they should also undergo RT-qPCR to verify that the target gene remains silenced at the same level from one experimental replicate to another.

6 Off-Target Effects

Since RNA silencing is driven by sequence identity between the dsRNA and target gene, it is important to ensure that the dsRNA sequence is specific to the gene of interest, and does not have potential for off-target silencing of non-target genes. That is, it is ideal to avoid designing dsRNA constructs that have close sequence identity to other non-target genes. If it is unavoidable, short stretches of similar sequence may not be an insurmountable problem in *Phytophthora* silencing experiments. Physically unlinked sequences with short regions of identity to the target sequence are typically not co-silenced [41, 47]. For example, the *P. infestans NIFS* gene, which shares sequence similarity with other *NIF* genes, escaped silencing when inverted repeat constructs targeting the *NIFC1* or *NIFC2* genes were used for silencing [41].

Phytophthora species lack genes encoding cytosine methyltransferases that can methylate DNA and transcriptionally silence genes, and no evidence for cytosine methylation has been found [13, 48]. In contrast, adenine N6-methylation can occur in *Phytophthora* genomes [49], but a link with RNA silencing mechanisms has not been established. Other chromatin modifications also occur at silenced loci, such as histone deacetylation and methylation, also leading to transcriptional inactivation through the formation of heterochromatin [48, 50]. In *P. infestans*, heterochromatin formation can extend outwards for short distances from silenced loci [41, 47]. This is a significant consideration in *Phytophthora* silencing experiments. *Phytophthora* genomes are organized into gene-poor "deserts" of repeat-rich sequence, and gene-dense "islands" [51]. The gene density in these islands means that neighboring genes can be separated by less than 1 kb. The formation of heterochromatin at a silenced locus therefore has potential to silence neighboring genes, potentially leading to confusing or incorrect phenotypes when analysing silenced transformants. There are no absolute rules about how distant a neighboring gene must be spaced to avoid the influence of heterochromatin from a silenced locus, although neighboring genes within 500 bp are often co-silenced [47]. The significance is that the effects of silencing on neighboring genes must be determined by RT-qPCR for each silencing experiment to be confident in the phenotypes observed. This can be less of a problem for effector coding genes which are frequently located in the gene-poor regions of *Phytophthora* genomes [51].

7 Considerations for Phenotyping Silenced Transformants

Phytophthora transformation results in insertions of the transgenes into gene-rich and gene-poor regions of the genome but with a disproportionate bias towards insertions near the 5′ end of genes [47]. As such, it is possible that these insertions can disrupt genome stability or regulation of other genes, leading to transformed lines with growth penalties or other unintended phenotypes [47]. Where the gene targeted for silencing may encode an essential or regulatory protein, it is possible that silenced lines may exhibit a growth penalty phenotype. It is therefore important to identify multiple silenced lines for each targeted gene, to be confident in the phenotypes observed.

Since *Phytophthora* sp. are plant pathogens, most studies that use silencing to knock down target gene expression also assess whether silenced lines have also been affected in their ability to infect and colonize plant tissues. If the growth of silenced lines in culture is impaired, this can raise a question of whether any reduction in pathogenicity is genuinely a result of silencing the target gene, or whether the impaired growth of the silenced lines allows plant tissues more opportunity to respond to and contain the infection (that is, the silenced lines are less fit due to non-specific effects from transformation). A further level of phenotypic information can be gained by carrying out infections of intact tissues and wounded tissues, to determine if silenced lines are impaired in the ability to successfully penetrate host surfaces and overcome the defences of the initially invaded cell, or if the ability to colonize host tissues is also impaired.

For silencing genes encoding secreted effectors, which would normally be upregulated during infection and not typically be expected to impact on pathogen growth in culture, it is important to use silenced lines that have no growth penalty resulting from transformation, so that the true impact of effector silencing on pathogenicity can be determined. That is, before proceeding to phenotype effector-silenced lines in infection assays, they should first be assessed for any growth penalty in culture on agar medium lacking selective antibiotics. This allows for the growth rate vs the non-transformed wild type to be determined, and not complicated by level of expression of the selectable marker gene used in transformation (for example *nptII*).

A further consideration when silencing genes encoding regulatory proteins (e.g., transcription factors) can be pleiotropic effects arising from the action of downstream genes/proteins also being affected by the silencing of the target gene. This can lead to phenotypic effects from silencing that become apparent in lifecycle stages that are not where the silenced target gene is normally maximally expressed.

Microscopy of specific lifecycle stages is commonly used to assess the impact of silencing specific genes. There are a wide variety of microscopic phenotypes that have been assessed, such as sporangia numbers, zoospore swimming patterns, zoospore aggregation, cyst germination, appressorium formation, and cell wall characteristics [19, 41, 46, 52, 53]. Other phenotypes that have been examined in culture have included osmotic stress, salt stress, reactive oxygen species stress, and laccase activity [54, 55]. During infection, trypan blue or aniline blue staining has often been used in the determination of microscopic examination of infection phenotypes arising from silencing [52, 56]. To date, transgene-mediated silencing of *Phytophthora* genes has not been carried out in a strain already expressing a fluorescent protein, or from a vector that co-expresses a fluorescent protein. This would enable more detailed confocal microscopy analysis of live plant tissues infected by silenced *Phytophthora* lines.

8 Conclusion

RNA silencing is an established tool for determining the roles of genes and the encoded proteins in *Phytophthora*. As for all functional genomics tools, RNA silencing has advantages and limitations. By careful consideration of how to apply this technique, it can be a powerful and highly informative tool to reveal the genetic components of *Phytophthora* pathogenicity and other lifecycle stages. There is scope for improvement of RNA silencing to increase its utility. For example, currently available silencing vectors utilize strong constitutive promotors to express inverted repeats to initiate silencing, which limits their utility when silencing essential genes; adaptation of existing inducible expression vectors [57] for silencing could expand the use of RNA silencing in *Phytophthora*. RNA silencing vectors that combine additional features such as fluorescent tags would also increase the phenotypic information that can be gained from silencing experiments. Together with techniques such as CRISPR-Cas, RNA silencing will continue to be valuable tool for characterizing the genes underpinning many aspects of *Phytophthora* biology.

Acknowledgements

SCW and LRJW acknowledge funding from the Scottish Government Rural and Environment Science and Analytical Services (RESAS) Division. RRV acknowledges financial support by the Swedish Research Council for Environment, Agricultural Sciences and Spatial Planning (FORMAS; grant numbers 2019-01316), Novo Nordisk Fonden (0074727), the Swedish Research Council (2019-04270) and Carl Tryggers Stiftelse för Vetenskaplig Forskning (CTS 20: 464).

References

1. Hannon GJ (2002) RNA interference. Nature 418:244–251
2. Bartel DP (2009) MicroRNAs: target recognition and regulatory functions. Cell 136:215–233
3. Ma X, Zuo Z, Shao W, Jin Y, Meng Y (2018) The expanding roles of Argonautes: RNA interference, splicing and beyond. Brief Funct Genomics 17:191–197
4. Zong J, Yao X, Yin J, Zhang D, Ma H (2009) Evolution of the RNA-dependent RNA polymerase (RdRP) genes: duplications and possible losses before and after the divergence of major eukaryotic groups. Gene 447:29–39
5. Maniar JM, Fire AZ (2011) EGO-1, a *C. elegans* RdRP, modulates gene expression via production of mRNA-templated short antisense RNAs. Curr Biol 21:449–459
6. Matzke MA, Mosher RA (2014) RNA-directed DNA methylation: an epigenetic pathway of increasing complexity. Nat Rev Genet 15: 394–408
7. Seroussi U, Li C, Sundby AE, Lee TL, Claycomb JM, Saltzman AL (2022) Mechanisms of epigenetic regulation by *C. elegans* nuclear RNA interference pathways. Semin Cell Dev Biol 127:142–154
8. Judelson HS, Whittaker SL (1995) Inactivation of transgenes in *Phytophthora infestans* is not associated with their deletion, methylation, or mutation. Curr Genet 28:571–579
9. van West P, Kamoun S, van 't Klooster JW, Govers F (1999) Internuclear gene silencing in *Phytophthora infestans*. Mol Cell 3:339–348
10. Fahlgren N, Bollmann SR, Kasschau KD, Cuperus JT, Press CM, Sullivan CM, Chapman EJ, Hoyer JS, Gilbert KB, Grünwald NJ, Carrington JC (2013) *Phytophthora* have distinct endogenous small RNA populations that include short interfering and microRNAs. PLoS One 8:e77181
11. Piombo E, Kelbessa BG, Sundararajan P, Whisson SC, Vetukuri RR, Dubey M (2023) RNA silencing proteins and small RNAs in oomycete plant pathogens and biocontrol agents. Front Microbiol 14:1076522
12. Qutob D, Chapman BP, Gijzen M (2013) Transgenerational gene silencing causes gain of virulence in a plant pathogen. Nat Commun 4:1349
13. Vetukuri RR, Avrova AO, Grenville-Briggs LJ, Van West P, Söderbom F, Savenkov EI, Whisson SC, Dixelius C (2011) Evidence for involvement of dicer-like, Argonaute and histone deacetylase proteins in gene silencing in *Phytophthora infestans*. Mol Plant Pathol 12: 772–785
14. Vetukuri RR, Åsman AKM, Tellgren-Roth C, Jahan SN, Reimegård J, Fogelqvist J, Savenkov E, Söderbom F, Avrova AO, Whisson SC, Dixelius C (2012) Evidence for small RNAs homologous to effector-encoding genes and transposable elements in the oomycete *Phytophthora infestans*. PLoS One 7: e51399
15. Åsman AKM, Fogelqvist J, Vetukuri RR, Dixelius C (2016) *Phytophthora infestans* Argonaute 1 binds micro RNA and small RNAs from effector genes and transposable elements. New Phytol 211:993–1007
16. Jia J, Lu W, Zhong C, Zhou R, Xu J, Liu W, Gou X, Wang Q, Yin J, Xu C, Shan W (2017) The 25–26 nt small RNAs in *Phytophthora parasitica* are associated with efficient silencing of homologous endogenous genes. Front Microbiol 8:773
17. Matson MEH, Liang Q, Lonardi S, Judelson HS (2022) Karyotype variation, spontaneous genome rearrangements affecting chemical insensitivity, and expression level polymorphisms in the plant pathogen *Phytophthora infestans* revealed using its first chromosome-scale assembly. PLoS Pathog 18:e1010869

18. Ah Fong AM, Judelson HS (2003) Cell cycle regulator Cdc14 is expressed during sporulation but not hyphal growth in the fungus-like oomycete *Phytophthora infestans*. Mol Microbiol 50:487–494
19. Blanco FA, Judelson HS (2005) A bZIP transcription factor from *Phytophthora* interacts with a protein kinase and is required for zoospore motility and plant infection. Mol Microbiol 56:638–648
20. Narayan RD, Blackman LM, Shan W, Hardham AR (2010) *Phytophthora nicotianae* transformants lacking dynein light chain 1 produce non-flagellate zoospores. Fungal Genet Biol 47:663–671
21. Sabbadin F, Urresti S, Henrissat B, Avrova AO, Welsh LRJ, Lindley PJ, Csukai M, Squires JN, Walton PH, Davies GJ, Bruce NC, Whisson SC, McQueen-Mason SJ (2021) Secreted pectin monooxygenases drive plant infection by pathogenic oomycetes. Science 373:774–779
22. Wang Y, Dou D, Wang X, Li A, Sheng Y, Hua C, Cheng B, Chen X, Zheng X, Wang Y (2009) The *PsCZF1* gene encoding a C2H2 zinc finger protein is required for growth, development and pathogenesis in *Phytophthora sojae*. Microb Pathog 47:78–86
23. Whisson SC, Avrova AO, van West P, Jones JT (2005) A method for double-stranded RNA--mediated transient gene silencing in *Phytophthora infestans*. Mol Plant Pathol 6:153–163
24. Cheng W, Lin M, Chu M, Xiang G, Guo J, Jiang Y, Guan D, He S (2022) RNAi-based gene silencing of RXLR effectors protects plants against the oomycete pathogen *Phytophthora capsici*. Mol Plant-Microbe Interact 35:440–449
25. Wang Z, Gao X, Zhong S, Li Y, Shi M, Zhang B, Zhang S, Shen H, Liu X (2022) Host-induced gene silencing of *PcCesA3* and *PcOSBP1* confers resistance to *Phytophthora capsici* in *Nicotiana benthamiana* through NbDCL3 and NbDCL4 processed small interfering RNAs. Int J Biol Macromol 222:1665–1675
26. Sanju S, Siddappa S, Thakur A, Shukla PK, Srivastava N, Pattanayak D, Sharma S, Singh BP (2015) Host-mediated gene silencing of a single effector gene from the potato pathogen *Phytophthora infestans* imparts partial resistance to late blight disease. Funct Integr Genomics 15:697–706
27. Jahan SN, Åsman AK, Corcoran P, Fogelqvist J, Vetukuri RR, Dixelius C (2015) Plant-mediated gene silencing restricts growth of the potato late blight pathogen *Phytophthora infestans*. J Exp Bot 66:2785–2794
28. Kalyandurg PB, Sundararajan P, Dubey M, Ghadamgahi F, Zahid MA, Whisson SC, Vetukuri RR (2021) Spray-induced gene silencing as a potential tool to control potato late blight disease. Phytopathology 111:2168–2175
29. Sharma S, Bairwa A, Tomar M, Kumar R, Bhardwaj V, Jeevalatha A, Bakade R, Salaria N, Thakur K, Singh BP, Chakrabarti SK (2022) Spraying of dsRNA molecules derived from *Phytophthora infestans*, along with nanoclay carriers as a proof of concept for developing novel protection strategy for potato late blight. Pest Manag Sci 78:3183–3192
30. Wang Y, Li M, Ying J, Shen J, Dou D, Yin M, Whisson SC, Birch PRJ, Yan S, Wang X (2023) High-efficiency green management of potato late blight by a self-assembled multicomponent nano-bioprotectant. Nat Commun 14:5622
31. Cvitanich C, Judelson HS (2003) Stable transformation of the oomycete, *Phytophthora infestans*, using microprojectile bombardment. Curr Genet 42:228–235
32. Huitema E, Smoker M, Kamoun S (2011) A straightforward protocol for electro-transformation of *Phytophthora capsici* zoospores. Methods Mol Biol 712:129–135
33. Sundararajan P, Kalyandurg PB, Liu Q, Chawade A, Whisson SC, Vetukuri RR (2022) Spray-induced gene silencing to study gene function in *Phytophthora*. Methods Mol Biol 2536:459–474
34. Welsh LRJ, Whisson SC (2023) Protoplast transformation of *Phytophthora* spp. Meth Mol Biol (in press; this volume)
35. Grenville-Briggs LJ, Anderson VL, Fugelstad J, Avrova AO, Bouzenzana J, Williams A, Wawra S, Whisson SC, Birch PR, Bulone V, van West P (2008) Cellulose synthesis in *Phytophthora infestans* is required for normal appressorium formation and successful infection of potato. Plant Cell 20:720–738
36. Mendoza CS, Findlay AC, Judelson H (2023) An LbCas12a variant and elevated incubation temperatures enhance the rate of gene editing in the oomycete *Phytophthora infestans*. Mol Plant-Microbe Interact 36:677. https://doi.org/10.1094/MPMI-05-23-0072-SC
37. Ah-Fong AMV, Bormann-Chung CA, Judelson HS (2008) Optimization of transgene-mediated silencing in *Phytophthora infestans* and its association with small-interfering RNAs. Fungal Genet Biol 45:1197–1205
38. Judelson HS, Tyler BM, Michelmore RW (1992) Regulatory sequences for expressing genes in oomycete fungi. Mol Gen Genet 234:138–146

39. Tani S, Yatzkan E, Judelson HS (2004) Multiple pathways regulate the induction of genes during zoosporogenesis in *Phytophthora infestans*. Mol Plant–Microbe Interact 17:330–337
40. Vetukuri RR, Tian Z, Avrova AO, Savenkov EI, Dixelius C, Whisson SC (2011) Silencing of the PiAvr3a effector-encoding gene from *Phytophthora infestans* by transcriptional fusion to a short interspersed element. Fungal Biol 115: 1225–1233
41. Judelson HS, Tani S (2007) Transgene-induced silencing of the zoosporogenesis-specific NIFC gene cluster of *Phytophthora infestans* involves chromatin alterations. Eukaryot Cell 6:1200–1209
42. Ah-Fong AM, Kim KS, Judelson HS (2017) RNA-seq of life stages of the oomycete *Phytophthora infestans* reveals dynamic changes in metabolic, signal transduction, and pathogenesis genes and a major role for calcium signaling in development. BMC Genomics 18:198
43. Wang S, Vetukuri RR, Kushwaha SK, Hedley PE, Morris J, Studholme DJ, Welsh LRJ, Boevink PC, Birch PRJ, Whisson SC (2021) Haustorium formation and a distinct biotrophic transcriptome characterize infection of *Nicotiana benthamiana* by the tree pathogen *Phytophthora kernoviae*. Mol Plant Pathol 22: 954–968
44. Evangelisti E, Gogleva A, Hainaux T, Doumane M, Tulin F, Quan C, Yunusov T, Floch K, Schornack S (2017) Time-resolved dual transcriptomics reveal early induced *Nicotiana benthamiana* root genes and conserved infection-promoting *Phytophthora palmivora* effectors. BMC Biol 15:39
45. Abrahamian M, Ah-Fong AM, Davis C, Andreeva K, Judelson HS (2016) Gene expression and silencing studies in *Phytophthora infestans* reveal infection-specific nutrient transporters and a role for the nitrate reductase pathway in plant pathogenesis. PLoS Pathog 12:e1006097
46. Leesutthiphonchai W, Judelson HS (2018) A MADS-box transcription factor regulates a central step in sporulation of the oomycete *Phytophthora infestans*. Mol Microbiol 110: 562–575
47. Vu AL, Leesutthiphonchai W, Ah-Fong AMV, Judelson HS (2019) Defining transgene insertion sites and off-target effects of homology-based gene silencing informs the application of functional genomics tools in *Phytophthora infestans*. Mol Plant-Microbe Interact 32: 915–927
48. van West P, Shepherd SJ, Walker CA, Li S, Appiah AA, Grenville-Briggs LJ, Govers F, Gow NAR (2008) Internuclear gene silencing in *Phytophthora infestans* is established through chromatin remodelling. Microbiology 154: 1482–1490
49. Chen H, Shu H, Wang L, Zhang F, Li X, Ochola SO, Mao F, Ma H, Ye W, Gu T, Jiang L, Wu Y, Wang Y, Kamoun S, Dong S (2018) *Phytophthora* methylomes are modulated by 6mA methyltransferases and associated with adaptive genome regions. Genome Biol 19:181
50. Wang L, Chen H, Li J, Shu H, Zhang X, Wang Y, Tyler BM, Dong S (2020) Effector gene silencing mediated by histone methylation underpins host adaptation in an oomycete plant pathogen. Nucleic Acids Res 48:1790–1799
51. Haas BJ, Kamoun S, Zody MC, Jiang RH, Handsaker RE, Cano LM, Grabherr M, Kodira CD, Raffaele S, Torto-Alalibo T, Bozkurt TO, Ah-Fong AM, Alvarado L, Anderson VL, Armstrong MR, Avrova A, Baxter L, Beynon J, Boevink PC, Bollmann SR, Bos JI, Bulone V, Cai G, Cakir C, Carrington JC, Chawner M, Conti L, Costanzo S, Ewan R, Fahlgren N, Fischbach MA, Fugelstad J, Gilroy EM, Gnerre S, Green PJ, Grenville-Briggs LJ, Griffith J, Grünwald NJ, Horn K, Horner NR, Hu CH, Huitema E, Jeong DH, Jones AM, Jones JD, Jones RW, Karlsson EK, Kunjeti SG, Lamour K, Liu Z, Ma L, Maclean D, Chibucos MC, McDonald H, McWalters J, Meijer HJ, Morgan W, Morris PF, Munro CA, O'Neill K, Ospina-Giraldo M, Pinzón A, Pritchard L, Ramsahoye B, Ren Q, Restrepo S, Roy S, Sadanandom A, Savidor A, Schornack S, Schwartz DC, Schumann UD, Schwessinger B, Seyer L, Sharpe T, Silvar C, Song J, Studholme DJ, Sykes S, Thines M, van de Vondervoort PJ, Phuntumart V, Wawra S, Weide R, Win J, Young C, Zhou S, Fry W, Meyers BC, van West P, Ristaino J, Govers F, Birch PR, Whisson SC, Judelson HS, Nusbaum C (2009) Genome sequence and analysis of the Irish potato famine pathogen *Phytophthora infestans*. Nature 461:393–398
52. Li A, Zhang M, Wang Y, Li D, Liu X, Tao K, Ye W, Wang Y (2014) PsMPK1, an SLT2-type mitogen-activated protein kinase, is required for hyphal growth, zoosporogenesis, cell wall integrity, and pathogenicity in *Phytophthora sojae*. Fungal Genet Biol 65:14–24
53. Latijnhouwers M, Ligterink W, Vleeshouwers VG, van West P, Govers F (2004) A G alpha subunit controls zoospore motility and virulence in the potato late blight pathogen *Phytophthora infestans*. Mol Microbiol 51:925–936

54. Gao J, Cao M, Ye W, Li H, Kong L, Zheng X, Wang Y (2015) PsMPK7, a stress-associated mitogen-activated protein kinase (MAPK) in *Phytophthora sojae*, is required for stress tolerance, reactive oxygenated species detoxification, cyst germination, sexual reproduction and infection of soybean. Mol Plant Pathol 16:61–70

55. Kong G, Chen Y, Deng Y, Feng D, Jiang L, Wan L, Li M, Jiang Z, Xi P (2020) The basic leucine zipper transcription factor PlBZP32 associated with the oxidative stress response is critical for pathogenicity of the lychee downy blight oomycete *Peronophythora litchii*. mSphere 5:e00261–e00220

56. Avrova AO, Boevink PC, Young V, Grenville-Briggs LJ, van West P, Birch PR, Whisson SC (2008) A novel *Phytophthora infestans* haustorium-specific membrane protein is required for infection of potato. Cell Microbiol 10:2271–2284

57. Gamboa-Meléndez H, Judelson HS (2015) Development of a bipartite ecdysone-responsive gene switch for the oomycete *Phytophthora infestans* and its use to manipulate transcription during axenic culture and plant infection. Mol Plant Pathol 16:83–91

Chapter 3

Protoplast Transformation of *Phytophthora* spp.

Lydia R. J. Welsh and Stephen C. Whisson

Abstract

At the core of assays to understand the role(s) of specific genes is the ability to stably transfer genes into *Phytophthora* through transformation. A key method for achieving this has been based on polyethylene glycol (PEG)/$CaCl_2$ transformation of protoplasts, but efficiency has often been low. Improving transformation efficiency is necessary for many applications, such as gene knockouts. Here we describe improvements through successive rounds of "mock" transformation, leading to improved efficiency in *Phytophthora infestans* and other species.

Key words *Phytophthora*, Transformation, Oomycete, Protoplasts, Antibiotic selection

1 Introduction

The genus *Phytophthora* encompasses some of the most damaging plant pathogens in agriculture and natural environments, such as *P. infestans* (potato and tomato late blight), *P. sojae* (soybean root rot), *P. ramorum,* and *P. cinnamomi* (tree pathogens) [1]. There has been an upsurge of genomics information for this genus in recent years, with genome and transcriptome data available for numerous species. Testing the role(s) of the genes identified within these resources requires techniques to introduce and express, or transform, versions of genes and foreign genes into *Phytophthora* species.

Transformation is the process of incorporating exogenous DNA into the cell of an organism. Several methods for transformation have been employed in *Phytophthora* spp.: electroporation of zoospores [2], *Agrobacterium*-mediated transformation [3], microprojectile bombardment [4] and protoplast-polyethylene glycol (PEG) /$CaCl_2$ [5]. Microprojectile bombardment requires specialized and expensive equipment, and many *Phytophthora* do not readily produce sufficient zoospores in vitro to enable electroporation to be carried out. Protoplast-PEG/$CaCl_2$ mediated

Monica L. Gerth and Rosie E. Bradshaw (eds.), *Phytophthora: Methods and Protocols*, Methods in Molecular Biology, vol. 2892, https://doi.org/10.1007/978-1-0716-4330-3_3,

transformation was the first method to be developed [5] and results in the insertion of multiple copies of the plasmid into the genome. In contrast, *Agrobacterium*-mediated transformation typically incorporates single copies of transforming DNA into the genome [3], which may result in lower transgene expression levels. For applications such as gene silencing in *P. infestans*, protoplast-PEG/$CaCl_2$ transformation yielded a two-fold increase in silenced transformants compared to zoospore electroporation and microprojectile bombardment, and transgene copy number was positively correlated with silencing [6]. Variations of the protoplast-PEG-$CaCl_2$ method have been used to transform numerous *Phytophthora* species, such as *P. infestans* [5, 7], *P. sojae* [8, 9], *P. capsici* [10], *P. cinnamomi* [11], *P. cactorum* [12], *P. palmivora* [13], *P. parasitica* [14], *P. ramorum* [15], *P. brassicae* [16], *P. litchii* [17] and *P. kernoviae* [18], demonstrating its versatility.

In the protoplast-PEG-$CaCl_2$ method, plasmid constructs with a selectable marker gene are typically transformed into *Phytophthora* protoplasts using a high concentration of the high molecular weight compound polyethylene glycol (PEG) and cationic liposomes containing DNA. Transformants are then recovered following protoplast regeneration on osmotically stabilized media containing a selective antibiotic. The resulting colonies can then be screened for expression of transgenes through microscopy, Western blot, or gene expression analysis.

Transformation of *Phytophthora* has been used for many purposes, for example, to overexpress avirulence genes [19], express fluorescent proteins and tag proteins with fluorescent markers [7], assess gene promoter activity [20, 21], silence genes [6, 12], and carry out targeted gene disruptions or knockouts using CRISPR-Cas [9, 22]. Expression of a cytoplasmic fluorescent protein can be used to visualize *Phytophthora* whilst infecting its living host [11]. Fluorescent tags are also used for protein localization through confocal microscopy [23]. The biological function of a protein can be investigated through epitope tagging, allowing Western blotting and co-immunoprecipitation to be carried out [24]. Mutated proteins can be expressed to further understand their function [19]. Silencing or knocking out a protein-coding gene can aid in determining its biological function by preventing its normal functional role, and then running phenotyping, transcriptomic, and proteomic assays [24, 25].

Vector choice depends on the application, but vector constructs must include an antibiotic-resistance gene for selecting transformants, oomycete-compatible promotor and terminator sequences, and a cloning site. The promoter and terminator sequences of the *Ham34* and *Hsp70* genes from the oomycete *Bremia lactucae* were shown to function well in *Phytophthora* compared to those from other eukaryotes [26]. More recently, ribosomal protein gene promotors have been used successfully [27]. Expression can also be driven by endogenous promoters to achieve stage-specific expression [21, 28], or inducible promotors to allow precise control of expression [29]. Antibiotic resistance genes for geneticin (G418; neomycin phosphotransferase; *npt*), hygromycin B (hygromycin phosphotransferase; *hpt*), streptomycin (streptomycin phosphotransferase), or gentamycin (N-acetyltransferase AAC(3)-1) have been described for selecting *Phytophthora* transformants [5, 30, 31]. Of the available selection agent/resistance gene combinations, geneticin (G418)/*npt* is the most commonly used. The pairing of the agrichemical oxathiapiprolin and its oxysterol-binding protein target has also been described as a potential selectable marker for *Phytophthora* transformation [32].

The efficiency of *Phytophthora* transformation can be influenced by the species, isolate, growth medium, and transformation construct. Optimization of growth medium is needed for each species to be transformed, both for initial culture growth and for regeneration of protoplasts. Sporangia can be collected from caducous *Phytophthora* species and used to grow a culture of young hyphae for protoplast generation. Hyphae of species that do not sporulate readily in culture, or with persistent spores, can be macerated briefly in a blender and used instead of sporangia as inoculum to provide young hyphae for protoplast production. Different enzyme preparations such as Novozym 234 [5] or Lysing Enzymes from *Trichoderma harzianum* [19] have been used with fungal cellulases for generating *Phytophthora* protoplasts from hyphae but are no longer commercially available. This protocol makes use of commercial enzyme preparations used in wine production, combined with a fungal cellulase, to produce protoplasts.

We have found that transformation efficiency can be increased by successive rounds of "mock" transformation. This is done by following the transformation protocol but omitting Lipofectin (used to form DNA-containing liposomes) and plasmid, then regenerating the protoplasts on growth medium without antibiotic selection. This then becomes the starter culture for transformation. After one initial round of "mock" transformation, a small aliquot of protoplasts can be regenerated from transformation experiments to repeat the "mock" transformation and, when repeated, continues to improve and maintain transformation efficiency. We have successfully transformed *P. infestans* (Clade 1), *P. pseudosyringae* (Clade 3), *P. rubi* (Clade 7), *P. fragariae* (Clade 7), *P. ramorum* (Clade 8), *P. kernoviae* (Clade 10), and *P. quercina* (Clade 12) using this protoplast-PEG/$CaCl_2$ protocol.

2 Materials

All water should be deionized and preferably 18 MΩ-cm at 25 °C. All reagents should be analytical or molecular biology grade. Store all solutions at room temperature unless otherwise stated.

1. **Agar plate culture**: Appropriate solid growth medium for the selected *Phytophthora* species (*see* **Note 1**).
2. **Lima bean extract**: Soak dried lima beans (also called butter beans; *Phaseolus lunatus*) overnight in deionized water at 4 °C. Drain and weigh 112 g of soaked beans, add to 750 mL deionized water and pulse blend for 5 s. Transfer to a 1 L bottle and autoclave (121 °C for 20 min). Once cool, sieve through muslin to remove lumps and freeze at −80 °C.

   Defrost extract on the day of making ALB broth. Once defrosted, remove any lumps by filtering through muslin. Centrifuge the extract at 10,000× *g* for 20 min and use the supernatant as lima bean extract in ALB broth (*see* **Note 2**).
3. **Trace elements**: In 400 mL water, dissolve 215 mg Fe(III) citrate, 150 mg $ZnSO_4.7H_2O$, 30 mg $CuSO_4$, 15 mg $MnSO_4.H_2O$, 10 mg H_3BO_4, 7 mg MoO_3. This will precipitate on storage; shake well before adding to ALB broth.
4. **Amended Lima Bean** (ALB) broth: Per liter: 2 g yeast extract, 1 g K_2HPO_4, 1 g KH_2PO_4, 3 g KNO_3, 0.5 g $MgSO_4$, 0.1 g $CaCl_2$, 5 g sorbitol, 5 g mannitol, 5 g dextrose, 2 mL trace elements (see Subheading 2, **step 3**), 250 mL lima bean extract (see Subheading 2, **step 2**). Pour ALB into 200 mL aliquots in bottles or flasks, and autoclave at 121 °C for 20 min.
5. **Filters**: Autoclavable filter units with 100 μm, 70 μm and 30 μm nylon mesh filters.
6. **Protoplasting enzymes:** Extralyse (Laffort) and cellulase from *Trichoderma viride* (Duchefa) or *Trichoderma reesei* (Merck) (*see* **Note 3**).
7. **Fry's Protoplasting Buffer (FPB)**: 0.4 M mannitol, 20 mM KCl, 10 mM $CaCl_2$, 20 mM 2-(morpholino)ethanesulfonic acid (MES), pH 5.7. Autoclave at 121 °C for 20 min.
8. **MT solution:** 1 M mannitol, 10 mM Tris-HCl, pH 7.5, 10 mM $CaCl_2$. Autoclave at 121 °C for 20 min.
9. **LBSM**: ALB broth with 20 g/L sucrose, 1 M mannitol. Autoclave at 121 °C for 20 min.
10. **PEG solution**: 50% PEG 3350, 10 mM $CaCl_2$, 10 mM Tris-HCl pH 7.5 (*see* **Note 4**).
11. **Plasmid DNA:** Maxi-plasmid preparations from *Escherichia coli* using a silica-based membrane/column kit. Completely

dry the plasmid DNA pellet under laminar flow to keep sterile, and dissolve in sterile molecular biology grade water. It is best to aim for higher concentrations of plasmid (>2 μg/μl). Typically, 200 mL of overnight *E. coli* culture with a high copy plasmid will yield at least 250 μg plasmid.

12. **Rye agar:** To make 1 L. Place 60 g of rye grain in a shallow tray and add enough sterile deionized water to cover the grain. Cover tightly and soak the rye grain for 42 h at room temperature in darkness. Filter the liquid through two thicknesses of muslin cloth into a flask and add 20 g sucrose to the flask. Transfer the soaked rye grain to a blender, add enough deionized water to cover the grain by approximately 2.5 cm, and macerate for 30 s. Incubate the macerated grain at 48.3 °C for 3 h in a water bath. Filter through two thicknesses of muslin cloth, squeezing gently to remove residual liquid. Discard the grain sediment. Combine the filtrate-sucrose solution from the soaked grain with the filtrate from the macerated grain. Adjust pH to 7.0 with 5 N NaOH and adjust to 1 L with deionized water. Dispense into bottles, add 1.5% agar and autoclave.
13. **100 mg/mL Antibiotic G418**, as a selective antibiotic (or other selective antibiotic to match the selectable marker in the transformation plasmid). Prepare in water and filter sterilize.
14. **100 mg/mL Ampicillin**. Prepare in water and filter sterilize. Used to prevent bacterial contamination (*see* **Note 5**).
15. **1% Pimaricin (Natamycin)** in 70% ethanol. Used to prevent fungal contamination (see **Note 5**).
16. **Lipofectin transfection reagent**, 1 mg/mL in water (Thermo Fisher Scientific).
17. 70% ethanol, 30% water.
18. Hemocytometer.
19. 20 mL syringes and 0.2 μm syringe filters.
20. 30 mL polystyrene universal tubes.
21. 150 mm and 90 mm Petri dishes.
22. Plastic spreaders.
23. Parafilm, or equivalent sealing film.
24. Metal forceps.
25. Scalpel.
26. Light microscope.
27. 50 mL centrifuge tubes, sterile.

3 Methods

With many steps involving open culture containers and transfer of solutions, sterile technique is vitally important in this protocol. Use 70% ethanol to decontaminate everything in the biosafety cabinet.

3.1 Determine Lethal Antibiotic Concentration Through a "Kill Curve" with the Appropriate Selective Antibiotic (see Note 6)

1. Pour agar with an increasing concentration of selective antibiotic into 90 mm Petri dishes. Start with no selective antibiotic and increase in 1 μg/mL increments.
2. Using a sterile scalpel, cut out 1 cm^2 pieces of an actively growing agar culture of *Phytophthora* and place one in the middle of each plate.
3. Grow in the dark at 18 °C for 12–14 d (*see* **Note** 7).
4. Measure the diameter of culture growth on each plate.
5. Plot the values on a graph of culture growth diameter vs antibiotic concentration.
6. Note the lowest concentration of selective antibiotic that results in no growth of *Phytophthora*. This is the lethal concentration of that selective antibiotic (*see* **Note 8**).

3.2 Prepare "Mock" Transformant

1. Inoculate five 150 mm Petri dishes of rye agar (50 μg/mL ampicillin, 0.001% pimaricin) with four agar plugs of wild-type *Phytophthora* culture (*see* **Note 1**). Grow in the dark at 18 °C for 12–14 d (*see* **Note** 7).
2. Check plates for any contamination before using as a starter culture. Discard any contaminated plates.
3. Pour 10–15 mL sterile deionized water onto a culture plate and use a sterile spreader to detach sporangia (*see* **Note 9**).
4. Use the spreader to carefully decant liquid onto another culture plate and collect sporangia by rubbing with the spreader.
5. Repeat for the rest of the culture plates, then decant the liquid into a filter unit fitted with a 70 μm nylon mesh filter. The sporangia will pass through the mesh filter.
6. Divide the sporangia suspension between two 200 mL bottles of ALB +50 μg/mL ampicillin.
7. Grow without shaking in the dark at 18 °C for 48 h (*see* **Note** 7).
8. In 20 mL FPB, dissolve 5 mg/mL Extralyse and 2 mg/mL cellulase. Syringe filter sterilize (0.2 μm) into a 50 mL centrifuge tube. Cover to protect from light and set aside.
9. Vortex the PEG solution to mix and set aside (*see* **Note 4**).

10. Filter the bottles of growing culture through 70 μm nylon mesh. The mycelia will be collected on the filter. Wash any ungerminated sporangia through with 100 mL sterile water.
11. Collect the mycelia from the filter with sterile forceps and transfer to the FPB-enzyme solution. Mix gently on a shaker platform (60–120 rpm) for 30–45 min at room temperature. Young hyphae will be digested by this point.
12. Pour the digested mycelia through a 100 μm nylon mesh filter, then a 30 μm nylon mesh filter. The protoplasts will pass through both of these filters. Transfer protoplasts to a 50 mL centrifuge tube. Centrifuge at 700× *g* for 4 min to pellet the protoplasts.
13. Syringe filter sterilize (0.2 μm) the PEG solution whilst the protoplasts are spinning.
14. Discard supernatant from Subheading 3.2, **step 12**. Gently resuspend the protoplasts in 10 mL FPB (*see* **Note 10**). Centrifuge again at 700× *g* for 4 min.
15. Discard supernatant. Resuspend the protoplasts in 5 mL FPB and slowly add 5 mL MT. Centrifuge again at 700× *g* for 4 min.
16. Discard supernatant. Resuspend the protoplasts in 10 mL MT. Centrifuge again at 700× *g* for 4 min.
17. Discard supernatant. Thoroughly resuspend the protoplast pellet in 1 mL MT, and transfer to a 30 mL polystyrene universal tube.
18. Allow to stand for 5 min, then slowly add 1 mL of PEG solution while rotating the tube at a 45° angle.
19. Keep rotating the tube for 30 s.
20. Allow to stand for 2 min, then mix once by inverting.
21. Allow to stand for 5 min.
22. Gently add 2 mL of LBSM and mix once by inversion.
23. Allow to stand for 2 min, then gently add 6 mL of LBSM and mix once by inversion.
24. Allow to stand for 3 min, then mix once more by inversion and pour into a 90 mm diameter Petri dish containing 12 mL LBSM with 50 μg/mL ampicillin.
25. Seal the dish with parafilm and allow protoplasts to regenerate in the dark for 48 h (*see* **Note 7**).
26. Transfer the regenerated protoplasts to a 50 mL centrifuge tube and centrifuge at 700× *g* for 5 min. An additional spin of 1000× *g* for 5 min can be carried out if mycelium has not pelleted.
27. Carefully pipette off the supernatant and resuspend the pellet in 2 mL LBSM.

28. Gently spread the regenerated protoplasts onto five 150 mm diameter Petri dishes of rye agar containing 50 μg/mL ampicillin and 0.001% pimaricin. Pipette approximately 500 μl of regenerated protoplasts onto each plate and push gently with a spreader to distribute. *P. infestans* will recover in 7 days, but recovery time for other species may be variable. Plates should be clear of any contamination.

3.3 Transformation

1. Inoculate 40 × 150 mm Petri dishes of rye agar (50 μg/mL ampicillin, 0.001% pimaricin) with four plugs of "mock" *Phytophthora* culture from Subheading 3.2, **step 28** (*see* **Note 1**). Grow in the dark at 18 °C for 12–14 days (*see* **Note 7**).
2. Check plates for any contamination before using as a starter culture. Discard any contaminated plates.
3. Pour 10–15 mL sterile deionized water onto a culture plate and use a sterile spreader to detach sporangia (*see* **Note 9**).
4. Use the spreader to carefully decant liquid onto another culture plate and collect more sporangia by rubbing with the spreader.
5. Repeat for a total of five culture plates, then decant the liquid into a filter unit fitted with a 70 μm nylon mesh filter. The sporangia will pass through the filter.
6. Repeat Subheading 3.3, **step 3–5**, for the remainder of the culture plates.
7. Distribute the sporangia suspension among 8 × 200 mL bottles of ALB +50 μg/mL ampicillin.
8. Grow without shaking in the dark at 18 °C for 48 h (*see* **Note 7**).
9. Prepare two 50 mL centrifuge tubes of FPB-enzyme solution, each containing 20 mL FPB, 5 mg/mL Extralyse and 2 mg/mL cellulase. Syringe filter sterilize (0.2 μm), cover to protect from light and set aside.
10. Prepare PEG solution, vortex to mix and set aside (*see* **Note 4**).
11. Filter four bottles of the growing culture through 70 μm nylon mesh. The mycelia will collect on the mesh. Pour the filtrate containing ungerminated sporangia back into the culture bottles to continue germinating. Wash any ungerminated sporangia through the filter with a 200 mL bottle of uninoculated ALB. Collect the mycelia from the filter with sterile forceps and transfer to the FPB-enzyme solution. Repeat for the rest of the culture bottles with another filter unit. Keep all 10 bottles of ALB to do further transformations on following days (*see* **Note 11**).
12. Mix gently on a shaker platform (60–120 rpm) for 30–45 min at room temperature. Young hyphae will be digested by this point.

13. Filter the digested mycelia through a 100 μm nylon mesh filter, then a 30 μm filter. The protoplasts will pass through both of these filters. Transfer the protoplasts to a 50 mL centrifuge tube. Centrifuge at 700× *g* for 4 min to pellet the protoplasts.
14. While the protoplasts are spinning, syringe filter sterilize the PEG solution. Also prepare the Lipofectin/plasmid mix (*see* **Note 12**): 50 μg plasmid in 57 μL sterile molecular biology grade water, mixed with 86 μL Lipofectin in a 30 mL polystyrene universal tube. Prepare one DNA-Lipofectin mixture for each centrifuge tube of protoplasts (*see* **Note 13**).
15. Discard the supernatant from Subheading 3.3, **step 13**. Gently resuspend the protoplasts in 10 mL FPB. Centrifuge again at 700× *g* for 4 min.
16. Discard supernatant. Gently resuspend the protoplasts in 5 mL FPB and slowly add 5 mL MT. Centrifuge again at 700× *g* for 4 min.
17. Discard supernatant. Gently resuspend the protoplasts in 10 mL MT. Centrifuge again at 700× *g* for 4 min.
18. Discard supernatant. Resuspend the protoplasts gently in 1 mL MT.
19. Use a hemocytometer to check the concentration of protoplasts at this stage. Expect between 1×10^6 and 1×10^7 protoplasts per mL.
20. Transfer approximately 50 μL of protoplasts from each centrifuge tube and pool in a sterile universal tube. Treat this as a "mock" transformation. Add the remainder of the protoplasts to the universal tubes containing the Lipofectin/plasmid mixtures and mix gently by rolling the tubes at a 45° angle for 30 s.
21. Allow to stand for 5 min.
22. Slowly add 1 mL of PEG solution whilst rotating the tube at a 45° angle.
23. Rotate the tube for a further 30 s.
24. Allow to stand for 2 min, then mix by inverting once.
25. Allow to stand for 5 min.
26. Gently add 2 mL of LBSM and mix once by inversion.
27. Allow to stand for 2 min.
28. Gently add 6 mL of LBSM and mix once by inversion.
29. Allow to stand for 3 min.
30. Mix once more by inversion and pour into a 90 mm diameter Petri dish containing 12 mL LBSM with 50 μg/mL ampicillin.
31. Seal the dish with Parafilm and allow protoplasts to regenerate in the dark for 48 h (*see* **Note 7**).

32. Transfer regenerated protoplasts to a 50 mL centrifuge tube and centrifuge at 700× *g* for 5 min. If mycelia have not pelleted, an additional spin of 1000× g for 5 min can be carried out.
33. Carefully pipette off the supernatant and resuspend the pellet in 2 mL LBSM.
34. Gently spread the regenerated protoplasts onto 5 × 150 mm diameter Petri dishes of rye agar; with 50 μg/mL ampicillin, 0.001% pimaricin and the lethal concentration of selective antibiotic determined from Subheading 3.1, **step 6**. For "mock" transformed protoplasts, omit the selective antibiotic and pipette approximately 500 μL of regenerated protoplasts onto each plate and push gently with a spreader to distribute.

3.4 Picking Transformed Colonies

1. Transformed colonies of *P. infestans* typically appear after 7–14 d, but the recovery time may vary with different strains and between species. Cut out colonies with a sterile scalpel as they appear and transfer to new agar plates containing the lethal dose of selective antibiotic (*see* **Note 14**).
2. Once transformants are actively growing on individual plates, transfer to higher doses of selective antibiotic for maintenance (*see* **Note 15**).

4 Notes

1. If using a *Phytophthora* species with caducous (easily detached) sporangia, choose a growth medium that yields high numbers of sporangia. For *P. infestans,* rye agar (see Subheading 2, **step 12**) is ideal for sporulation. The incubation temperature and time should be optimized for the species being used for transformation. Avoid using old agar plate cultures for transformation, as the sporangia do not germinate well.
2. ALB should be clear with no or little sediment; it needs to pass through filters without clogging. The freezing process causes additional polysaccharides and proteins to precipitate, which then do not precipitate later upon autoclaving when included in ALB medium.
3. The ratio and concentration of protoplasting enzymes used in this protocol are effective for *P. infestans* and several other *Phytophthora* species but may need to be optimized if hyphae are recalcitrant to digestion.
4. PEG3350: Batches of 1 M Tris-HCl (pH 7.5) and 100 mM $CaCl_2$ can be prepared and frozen in small aliquots to use on the day of transformation. Mix 2.5 g of PEG3350, 50 μL of 1 M Tris-HCl (pH 7.5), 500 μL of 100 mM $CaCl_2$, and water

to a final volume of 5 mL to obtain a final concentration of 10 mM each. Always prepare this fresh on the day of transformation.

5. Other antibiotics used to suppress contamination include vancomycin (50 μg/mL) and rifampicin (10 μg/mL). Fungicides such as nystatin (25 μg/mL), pentachloronitrobenzene (PCNB; 25 μg/mL), or hymexazol (50 μg/mL) can also be used but may slow *Phytophthora* growth. Antibiotics are not a replacement for sterile technique.
6. Always carry out a "kill curve" whenever a new isolate is being used for transformation, even if the same species.
7. Timings and conditions stated are optimal for *P. infestans* but may need optimization for other *Phytophthora* species.
8. For example, an isolate of *P. infestans* has slow growth on 1–2 μg/mL G418 but no growth at 3 μg/mL. Thus 3 μg/mL is a good choice as a low lethal dose for initial selection.
9. For species that do not sporulate readily in culture, use a rich solid medium to encourage healthy hyphal growth. Use mycelial fragments scraped from five 150 mm Petri dishes as inoculum for two bottles of ALB and grow until mycelia are visibly clumping. Macerate mycelia in a sterile blender for 10 s under sterile conditions and aliquot the mix into ten bottles of fresh ALB. Grow for no more than two days; old cultures do not digest well with protoplasting enzymes and very few viable protoplasts are recovered.
10. Protoplasts are fragile. Treat them gently. Protoplasts tend to form clumps that are difficult to resuspend after centrifuging. Resuspend with a wide-bore pipette tip (such as for a 5 mL pipette) and let the liquid run down the side of the centrifuge tube rather than directing a jet of liquid at the protoplast pellet. Gently pipette the liquid up and down to dislodge the pellet. Do not vortex.
11. More transformations can be carried out on subsequent days. Spore germination is not synchronous and for *P. infestans* it can be possible to perform transformation up to six times from the same set of culture bottles. Remember that young hyphae produce more viable protoplasts, and ensure cultures are clean when starting. If using blended mycelia instead of spores as inoculum, typically only one or two days of transformations will be possible.
12. Lipofectin may not be needed. If Lipofectin is not used, increase the concentration of $CaCl_2$ to 25 mM in both MT solution and PEG preparation.
13. Exceeding 50 μg of plasmid per transformation does not improve efficiency. The DNA-Lipofectin mixture may become slightly cloudy; this is normal.

14. If unsuccessful, try a different isolate. Transformation efficiency can vary between isolates, although fast growing isolates are easier to work with.
15. Maintenance: Transformants of *P. infestans* and many other species are routinely maintained on rye agar containing 10 μg/mL G418.

Acknowledgements

LRJW and SCW acknowledge funding from the Scottish Government Rural and Environment Science and Analytical Services (RESAS) Division.

References

1. Kamoun S, Furzer O, Jones JD, Judelson HS, Ali GS, Dalio RJ, Roy SG, Schena L, Zambounis A, Panabières F, Cahill D, Ruocco M, Figueiredo A, Chen XR, Hulvey J, Stam R, Lamour K, Gijzen M, Tyler BM, Grünwald NJ, Mukhtar MS, Tomé DF, Tör M, Van Den Ackerveken G, McDowell J, Daayf F, Fry WE, Lindqvist-Kreuze H, Meijer HJ, Petre B, Ristaino J, Yoshida K, Birch PR, Govers F (2015) The top 10 oomycete pathogens in molecular plant pathology. Mol Plant Pathol 16:413–434
2. Huitema E, Smoker M, Kamoun S (2011) A straightforward protocol for electro-transformation of *Phytophthora capsici* zoospores. Methods Mol Biol 712:129–135
3. Vijn I, Govers F (2003) *Agrobacterium tumefaciens* mediated transformation of the oomycete plant pathogen *Phytophthora infestans*. Mol Plant Pathol 4:459–467
4. Cvitanich C, Judelson HS (2003) Stable transformation of the oomycete, *Phytophthora infestans*, using microprojectile bombardment. Curr Genet 42:228–235
5. Judelson HS, Tyler BM, Michelmore RW (1991) Transformation of the oomycete pathogen, *Phytophthora infestans*. Mol Plant–Microbe Interact 4:602–607
6. Ah-Fong AMV, Bormann-Chung CA, Judelson HS (2008) Optimization of transgene-mediated silencing in *Phytophthora infestans* and its association with small-interfering RNAs. Fungal Genet Biol 45:1197–1205
7. Ah-Fong AM, Judelson HS (2011) Vectors for fluorescent protein tagging in *Phytophthora*: tools for functional genomics and cell biology. Fungal Biol 115:882–890
8. Judelson HS, Coffey MD, Arredondo FR, Tyler BM (1993) Transformation of the oomycete pathogen *Phytophthora megasperma f. sp. glycinea* occurs by DNA integration into single or multiple chromosomes. Curr Genet 23: 211–218
9. Cao J, Qiu M, Ye W, Wang Y (2022) *Phytophthora sojae* transformation based on the CRISPR/Cas9 system. Bio Protoc 12:e4352
10. Wang Z, Tyler BM, Liu X (2018) Protocol of *Phytophthora capsici* transformation using the CRISPR-Cas9 system. Methods Mol Biol 1848:265–274
11. Dai T, Xu Y, Yang X, Jiao B, Qiu M, Xue J, Arredondo F, Tyler BM (2021) An improved transformation system for *Phytophthora cinnamomi* using green fluorescent protein. Front Microbiol 12:682754
12. Chen XR, Li YP, Li QY, Xing YP, Liu BB, Tong YH, Xu JY (2016) SCR96, a small cysteine-rich secretory protein of *Phytophthora cactorum*, can trigger cell death in the *Solanaceae* and is important for pathogenicity and oxidative stress tolerance. Mol Plant Pathol 17:577–587
13. van West P, Reid B, Campbell TA, Sandrock RW, Fry WE, Kamoun S, Gow NA (1999) Green fluorescent protein (GFP) as a reporter gene for the plant pathogenic oomycete *Phytophthora palmivora*. FEMS Microbiol Lett 178:71–80
14. Bottin A, Larche L, Villalba F, Gaulin E, Esquerré-Tugayé MT, Rickauer M (1999) Green fluorescent protein (GFP) as gene expression reporter and vital marker for studying development and microbe-plant interaction in the tobacco pathogen *Phytophthora parasitica var. nicotianae*. FEMS Microbiol Lett 176:51–56

15. Riedel M, Calmin G, Belbahri L, Lefort F, Götz M, Wagner S, Werres S (2009) Green fluorescent protein (GFP) as a reporter gene for the plant pathogenic oomycete *Phytophthora ramorum*. J Eukaryot Microbiol 56: 130–135
16. Si-Ammour A, Mauch-Mani B, Mauch F (2003) Quantification of induced resistance against *Phytophthora* species expressing GFP as a vital marker: beta-aminobutyric acid but not BTH protects potato and *Arabidopsis* from infection. Mol Plant Pathol 4:237–248
17. Situ J, Jiang L, Fan X, Yang W, Li W, Xi P, Deng Y, Kong G, Jiang Z (2020) An RXLR effector PlAvh142 from *Peronophythora litchii* triggers plant cell death and contributes to virulence. Mol Plant Pathol 21:415–428
18. Wang S, Vetukuri RR, Kushwaha SK, Hedley PE, Morris J, Studholme DJ, Welsh LRJ, Boevink PC, Birch PRJ, Whisson SC (2021) Haustorium formation and a distinct biotrophic transcriptome characterize infection of *Nicotiana benthamiana* by the tree pathogen *Phytophthora kernoviae*. Mol Plant Pathol 22: 954–968
19. Whisson SC, Boevink PC, Moleleki L, Avrova AO, Morales JG, Gilroy EM, Armstrong MR, Grouffaud S, van West P, Chapman S, Hein I, Toth IK, Pritchard L, Birch PRJ (2007) A translocation signal for delivery of oomycete effector proteins into host plant cells. Nature 450:115–118
20. Tani S, Judelson HS (2006) Activation of zoosporogenesis-specific genes in *Phytophthora infestans* involves a 7-nucleotide promoter motif and cold-induced membrane rigidity. Eukaryot Cell 5:745–752
21. Avrova AO, Boevink PC, Young V, Grenville-Briggs LJ, van West P, Birch PR, Whisson SC (2008) A novel *Phytophthora infestans* haustorium-specific membrane protein is required for infection of potato. Cell Microbiol 10:2271–2284
22. Fang Y, Tyler BM (2016) Efficient disruption and replacement of an effector gene in the oomycete *Phytophthora sojae* using CRISPR/Cas9. Mol Plant Pathol 17:127–139
23. Ah-Fong AMV, Kagda M, Judelson HS (2018) Illuminating Phytophthora biology with fluorescent protein tags. Methods Mol Biol 1848: 119–129
24. Qiu M, Li Y, Zhang X, Xuan M, Zhang B, Ye W, Zheng X, Govers F, Wang Y (2020) G protein α subunit suppresses sporangium formation through a serine/threonine protein kinase in *Phytophthora sojae*. PLoS Pathog 16: e1008138
25. Leesutthiphonchai W, Judelson HS (2018) A MADS-box transcription factor regulates a central step in sporulation of the oomycete Phytophthora infestans. Mol Microbiol 110: 562–575
26. Judelson HS, Tyler BM, Michelmore RW (1992) Regulatory sequences for expressing genes in oomycete fungi. Mol Gen Genet 234:138–146
27. Poidevin L, Andreeva K, Khachatoorian C, Judelson HS (2015) Comparisons of ribosomal protein gene promoters indicate superiority of heterologous regulatory sequences for expressing transgenes in *Phytophthora infestans*. PLoS One 10:e0145612
28. Xiang Q, Kim KS, Roy S, Judelson HS (2009) A motif within a complex promoter from the oomycete *Phytophthora infestans* determines transcription during an intermediate stage of sporulation. Fungal Genet Biol 46:400–409
29. Gamboa-Meléndez H, Judelson HS (2015) Development of a bipartite ecdysone-responsive gene switch for the oomycete *Phytophthora infestans* and its use to manipulate transcription during axenic culture and plant infection. Mol Plant Pathol 16:83–91
30. Judelson HS (1993) Intermolecular ligation mediates efficient cotransformation in *Phytophthora infestans*. Mol Gen Genet 239:241–250
31. Evangelisti E, Yunusov T, Shenhav L, Schornack S (2019) *N*-acetyltransferase AAC(3)-I confers gentamicin resistance to *Phytophthora palmivora* and *Phytophthora infestans*. BMC Microbiol 19:265
32. Wang W, Xue Z, Miao J, Cai M, Zhang C, Li T, Zhang B, Tyler BM, Liu X (2019) PcMuORP1, an Oxathiapiprolin-resistance gene, functions as a novel selection marker for *Phytophthora* transformation and CRISPR/Cas9 mediated genome editing. Front Microbiol 10:2402

Chapter 4

Gene Editing and Protein Tagging in the Oomycete *Phytophthora infestans* Using CRISPR-Cas12a

Carl S. Mendoza, Audrey M. V. Ah-Fong, and Howard S. Judelson

Abstract

Molecular genetic tools such as CRISPR-Cas gene editing systems are invaluable for understanding gene and protein function and revealing the details of a pathogen's life and disease cycles. Here we present protocols for genome editing in *Phytophthora infestans*, an oomycete with global importance as a pathogen of potato and tomato. Using a vector system that expresses variants of Cas12a from *Lachnospiraceae bacterium* and its guide RNA from a unified transcript, we first present a method for editing genes through the non-homologous end-joining (NHEJ) pathway. We then describe an application of homology-directed repair (HDR), in which Cas12a is used to fuse a protein-coding gene with a fluorescent or epitope tag. Both methods should be adaptable to many oomycetes other than *P. infestans*.

Key words Oomycete, Gene editing, CRISPR, Transformation, Fluorescent reporter

1 Introduction

Methods for modifying genomes derived from bacterial CRISPR systems are becoming indispensable for research in many biological systems [1]. CRISPR is already accelerating studies of several members of the oomycete genus *Phytophthora* [2–5], which includes some of the world's most destructive plant pathogens [6]. For example, the focus of this chapter is the globally significant potato and tomato pathogen *Phytophthora infestans* [7]. In combination with genome sequence and transcriptome data [8, 9], and other tools for genome manipulation such as DNA-mediated transformation, homology-based gene silencing, and systems for expressing fluorescently labeled proteins [10–13], the expanding toolkit for *Phytophthora* holds promise for illuminating its biology and helping to identify strategies for controlling plant disease.

The CRISPR tools used most frequently for studying bacteria and eukaryotes employ either the Cas9 or Cas12a (Cpf1) proteins [1]. Guided by RNA sequences targeting a genomic locus, both

Monica L. Gerth and Rosie E. Bradshaw (eds.), *Phytophthora: Methods and Protocols*, Methods in Molecular Biology, vol. 2892, https://doi.org/10.1007/978-1-0716-4330-3_4,

cause double-stranded breaks in the host chromosome. These are repaired by the host's non-homologous end-joining (NHEJ) pathway or, in the presence of a template, by homology-directed repair (HDR) [14]. Repair by NHEJ usually involves the removal or addition of several bases, often leading to a change in reading frame and a defective protein. The cellular role of the affected gene can then be assessed through phenotypic analysis. While HDR can also be used to knock out expression of a functional protein, its other applications include gene replacement, promoter modification, tagging genes with an epitope or fluorescent marker, or inserting DNA at a "safe harbor" to avoid affecting endogenous genes or obtaining consistent expression of a reporter or other transgene [15–17].

Cas9 and Cas12a differ in several important aspects. Each involves a protospacer adjacent motif (PAM), which is a short DNA sequence found near the DNA site that is targeted for cleavage. However, the PAM for the commonly used form of Cas9 is 5′-NGG while for Cas12a this is 5′-TTTV [18]. Another difference is that Cas12a, particularly from *Lachnospiraceae bacterium*, retains activity at lower temperatures to a greater extent than Cas9 [19]. This makes it attractive for use in ectothermic animals, plants, and many species of *Phytophthora*. Another difference is that CRISPR RNA precursors used for editing with Cas12a are much shorter than those employed by Cas9, since Cas12a can process its precursor directly without the trans-activating RNA scaffold required by Cas9 [20]. This makes synthesizing a Cas12a guide RNA (gRNA, also known as crRNA) less expensive and facilitates the construction of multiplex editing arrays [21]. Cas12a is also reported to generate larger deletions, which may increase the rate of loss-of-function events, and to have fewer-off target effects [22]. Despite these advantages of Cas12a, having both systems could be useful since their distinct PAM sequences increase the number of potentially editable sites.

Initial attempts to employ CRISPR-Cas9 in *P. infestans* failed due to the presumed toxicity of the enzyme, a situation which is not unique to oomycetes [23]. Instead, we developed a successful CRISPR-Cas12a system [3]. Unlike the Cas9 method deployed in other species of *Phytophthora* in which the editing protein and gRNA are encoded by separate genes [2], ours produces the nuclease and guide RNA in a single transcript to increase the likelihood that both would be expressed in a transformant. Another difference is that we make use of the innate RNase activity of Cas12a to form the mature guide RNA, instead of ribozyme sequences which appeared to function less effectively in *P. infestans* [3].

This chapter provides protocols for two distinct applications of Cas12a. First, we describe gene editing through NHEJ. Second, we illustrate the use of HDR to tag a protein, by expressing Cas12a and a guide RNA in the presence of a repair template encoding a fluorescent protein. The resulting fusion can be visualized by microscopy to investigate the function of the protein or other aspects of cellular dynamics. While our protocols focus on *P. infestans,* they should be adaptable to other members of the *Phytophthora* genus or close relatives such as *Pythium* species.

2 Materials

2.1 Vector Construction

1. Cas12a editing vector (pSTU1 or pSTUC1; *see* **Note 1**).
2. Upper and lower-strand oligonucleotides for constructing cassettes encoding gRNAs (see Subheading 3.1 for design instructions).
3. 42 °C heating block or water bath.
4. Microcentrifuge.
5. Thermal cycler.
6. Chemically competent (or electrocompetent) *Escherichia coli,* such as strain DH5α.
7. Lysogeny Broth agar (LB): 1.0% NaCl, 1.0% tryptone, 0.5% yeast extract, 1.5% agar.
8. Terrific Broth (TB): Combine 24 g yeast extract, 12 g tryptone, and 4 g glycerol in water to a total volume of 900 mL. Separately, mix 2.3 g KH2PO4 with 16.4 g M K2HPO4 in water to a total volume of 100 mL. Autoclave both solutions separately and then combine.
9. Super Optimal Catabolite broth (SOC): 2% tryptone, 0.5% yeast extract, 0.05% NaCl.
10. T4 DNA ligase and 10× buffer for T4 DNA ligase (including ATP).
11. 0.5 M NaCl.
12. BsaI restriction enzyme and buffer (see **Note 2**).
13. Small- and large-scale plasmid purification kits.
14. Primer 3′HAM34-R (5′-CGCCATCACCGATTCGTC- 3′).
15. For HDR, a general vector such as pBluescript for cloning the donor DNA.

2.2 Transformation of Protoplasts

1. Room temperature centrifuge with swinging-bucket rotor.
2. Orbital shaker capable of slow speeds (<50 rpm).
3. Hemocytometer.
4. Syringe filters, preferably low protein-binding with prefilter.
5. 50–70 μm nylon mesh or cell strainer.
6. Sterile glassware, including 1 L glass flasks and 400 mL beakers.

7. Forceps, preferably with an angled tip.
8. Sterile spreaders, either disposable plastic or made from a bent glass rod.
9. A well-sporulating diploid strain of *Phytophthora infestans*.
10. Rye-sucrose broth: For 1 L, soak 60 g clean rye seeds (preferably organic) in 0.5 L water at room temperature for 24 h. The mixture should smell "sweet" and become just slightly acidic, with a pH of about 6.0. Then, break the grains into smaller pieces (about 1/4 the size of a whole grain) using a blender, and autoclave for 30 min. After removing large grain pieces using a colander, add water to 1 L and 20 g sucrose. We recommend storing frozen aliquots which are sterilized before use.
11. Amended lima bean (ALBA) media: 5 g/L sorbitol, 5 g/L mannitol, 5 g/L dextrose, 3 g/L KNO_3, 1 g/L K_2HPO_4, 1 g/L KH_2PO_4, 0.5 g/L $MgSO_4$, 0.1 g/L $CaCl_2$, 2 g/L yeast extract, 2 mL vitamin stock, 2 mL trace elements, and 250 mL lima bean extract (prepared by blending and then autoclaving 280 g/L lima beans in water, then clarified by centrifugation). The vitamin stock (stored in the dark) contains 0.2 mg biotin, 0.2 mg folic acid, 12 mg I-inositol, 60 mg nicotinic acid, 18 mg pyridoxine-HCl, 15 mg riboflavin, 38 mg thiamine-HCl, and 50 mL coconut milk in 300 mL water. The trace element stock contains 215 mg $FeC_6H_5O_7.3H_2O$, 150 mg $ZnSO_4.7H_2O$, 30 mg $CuSO_4.5H_2O$, 10 mg $MnSO_4.H_2O$, 10 mg H_3BO_3, and 7 mg MoO_3 in 400 ml. This medium is stored at 4 °C. If a precipitate is seen in the ALBA after refrigeration, clarify by centrifugation for 15 min at 5000 × *g*.
12. Protoplasting enzymes: Cellulase and β-glucanase (*see* **Note 3**).
13. Protoplasting buffer (FPB): 0.4 M mannitol, 20 mM KCl, 20 mM MES pH 5.7, 10 mM $CaCl_2$.
14. Mannitol-Tris (MT): 1 M mannitol, 10 mM Tris pH 7.5.
15. Mannitol-Tris-Calcium (MTC): MT containing 10 mM $CaCl_2$.
16. Polyethylene glycol mix: 50% PEG-MW 3350 (Millipore-Sigma) containing 25 mM $CaCI_2$ and 10 mM Tris 7.5.
17. Lipofectin (Thermo Fisher Scientific; optional, *see* **Note 4**).
18. Protoplast regeneration media: rye-sucrose broth with 1 M mannitol, clarified by centrifugation for 15 min at 5000× *g*.
19. Selective media: rye-sucrose containing 1.5% agar and 8 μg/mL G-418 (*see* **Note 5**).

2.3 Detecting Editing

1. Boiling water bath.
2. Thermal cycler.
3. DNA extraction buffer: 0.2 M Tris pH 8.5, 0.25 M NaCl, 25 mM EDTA, 2% sodium dodecyl sulfate.
4. 1:1 phenol:chloroform.
5. 100% isopropanol.
6. 70% ethanol.
7. Low TE: 10 mM Tris pH 7.9, 0.1 mM EDTA.
8. Polymerase chain reaction (PCR) reagents: This can be purchased as a premade 2× mix or assembled from individual components. A 1× reaction would contain 50 mM KCl, 10 mM Tris-Cl pH 8.3, 1 mM $MgCl_2$, 200 μM deoxynucleoside triphosphate (dNTP) stock, and 1 unit *Taq* polymerase per 25 μL, plus primers.
9. PCR primers to screen for editing. These are usually designed to bind about 75 nt on either side of the site targeted for editing.
10. Kits or reagents for cleaning up samples prior to sequencing (*see* **Note 6**).

3 Methods

3.1 Designing Guide RNAs for Editing by NHEJ

1. We recommend using the browser-based programs EuPaGDT or DeepCpf1 to identify the gRNA [24, 25]. For EuPaGDT, select the Cpf1/Cas12a nuclease option, TTTV for the PAM, and 23 nt for the guide length. V indicates A, C, or G, but not T. Note that the gRNA will include the 23 nt downstream (3′ end) of the PAM, but not the TTTV itself (Fig. 1a). When evaluating gRNA candidates, they should have 35 to 65% GC and little to no secondary structure with no more than 3 consecutive paired bases [1]. For knocking out protein function, gRNAs close to the 5′ end of the coding sequences or upstream of a functional domain are preferred to maximize the likelihood that editing will prevent formation of an active protein.
2. If possible, select several gRNAs to maximize the probability of editing. These could be tested in separate plasmids, but we usually construct arrays of two to three gRNAs in a single plasmid. We prefer to select targets within 500 nt of each other, so that a single set of primers can be used for PCR and sequence analysis of each potential editing site.
3. Consider using chromatin accessibility data to prioritize gRNA candidates, since studies in other taxa have indicated that nucleosome-free regions are more editable [26].

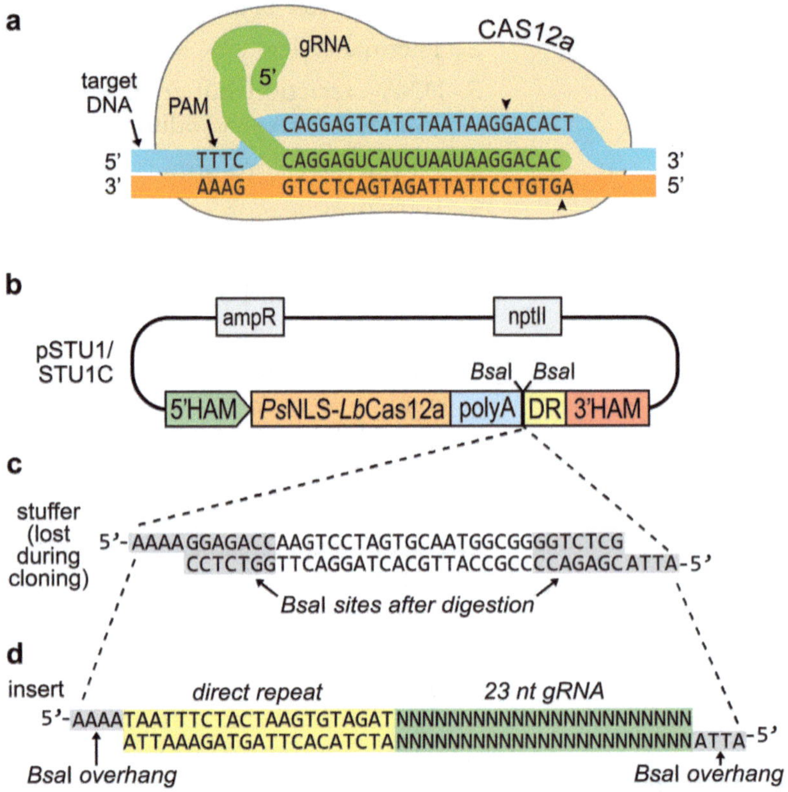

Fig. 1 (**a**) Target digestion by Cas12a. Labeled are a representative gRNA and PAM in host DNA. It should be noted that the orientation of the PAM relative to the gRNA is different for Cas9 and Cas12a. Arrowheads mark the staggered cuts made by Cas12a; these sites are approximations, since Cas12 often cuts at several positions [38]. (**b**) Schematic of Cas12a plasmids. Shown are the *Ham34* transcriptional promoter (5′HAM) and terminator (3′HAM) sequences, Cas12a sequences from *L. bacterium* fused to an oomycete nuclear localization motif (PsNLS), a 75 nt polyA region, and the Cas12a direct repeat. The ampicillin and G-418 (nptII) resistance genes used for selection in *E. coli* and *Phytophthora* are also illustrated. (**c**) Region of the plasmid liberated by *Bsa*I digestion of the pSTU plasmids. (**d**) Fragment inserted into the plasmids. This contains two *Bsa*I overhangs, the Cas12a direct repeat, and the 23 nt gRNA; the series of Ns will be changed based on the target sequence

4. It is preferable for gRNAs to match exactly their chromosomal targets, so check for potential polymorphisms between alleles within your strain of *Phytophthora* that might impair Cas12a activity or confound assays for editing (*see* **Note 7**).

3.2 Constructing the Editing Vector

1. Design the insert for the editing vector (Fig. 1b). Each gRNA should be preceded by a Cas12a direct repeat (DR), and the entire fragment should contain *Bsa*I overhangs at each end. If using an gRNA array, a DR must be placed between each (e.g.,

forming *Bsa*I–DR–gRNA1–DR–gRNA2–*Bsa*I). Adding a DR 3′ to the gRNA is unnecessary as one is within the backbone of the pSTU plasmids. While the insert for cloning can be designed to include whole *Bsa*I sites at each terminus, we find it more convenient to design two oligonucleotides which when annealed will have single-stranded tails compatible with the left and right *Bsa*I sites of the vector (TTTT and TAAT, respectively; Fig. 1d).

2. Order both the upper and lower strands from a commercial provider, and resuspend each in TE buffer to 100 μM.
3. Anneal the two oligonucleotides. First, combine 10 μL of each in a single tube and add 1 μL of 0.5 M NaCl to 0.05 M. Then, heat at 95 °C for 2 min, and cool to 4 °C over about 45 min. This can be done in a thermal cycler with a ramp program, or by heating the tube in a heat block which is then placed on the benchtop to cool to room temperature, followed by storage at 4 °C.
4. Digest the plasmid with tenfold excess of *Bsa*I. Following digestion, we recommend gel purification of the linearized plasmid to reduce any persisting undigested plasmid and to eliminate the stuffer fragment (Fig. 1c).
5. Mix the plasmid and insert in a 1:4 molar ratio and ligate. We normally perform ligations containing 1 μL of 10× T4 DNA ligase buffer with ATP, 1 μL (400 units) of T4 DNA ligase, a total of 100 ng of plasmid and insert, and water to 20 μL. This translates to about 98 ng of plasmid and 2 ng of insert for an insert with two gRNAs. Incubate overnight at 4 to 12 °C.
6. Mix 1 to 10 μL of the ligation reaction with 50 μL of competent *E. coli* cells (e.g., DH5α), incubate on ice for 30 min, and heat at 42 °C for 90 s (or 37 °C for 3 min). Then, add 400 μL of SOC or LB broth and place in a shaking incubator at 37 °C for 1 h. Plate aliquots on LB agar containing 100 μg/mL ampicillin or carbenicillin, and incubate 14–18 h at 37 °C until colonies are observed.
7. Identify clones containing the gRNA. Since the cloning procedure is highly efficient, we normally prepare plasmid DNA from three colonies and submit them for Sanger sequencing using primer 3′HAM34-R. This binds the *3′ Ham34* terminator allowing the sequence to read across the cloning site. Another option is to check the plasmids prior to sequencing for the absence of a *Bsa*I site, which should be lost upon cloning.
8. Prepare DNA for *Phytophthora* transformation using a scaled-up plasmid preparation method. In our hands, about 100 μg of DNA is typically more than sufficient for obtaining enough edited transformants, but this depends on the editing rate which varies depending on the gene and guide RNA (*see* **Note 8**).

3.3 Polyethylene Glycol-Mediated Protoplast Transformation

1. Prepare *P. infestans* cultures for protoplast transformation (*see* **Note 9**). Harvest sporangia from a 7- to 14-day-old stock culture by adding 10 mL of water followed by gentle rubbing with a spreader rod. Pipette off the liquid and count the sporangia with a hemocytometer. Spread about 5×10^4 sporangia on each of ten 150-mm plates of fresh rye-sucrose agar to establish the pre-inoculum culture. After allowing the liquid to absorb, invert the plates and grow for 7–9 d at 18 °C.
2. Isolate sporangia from the 150-mm plates by adding to each 15 mL of sterile H_2O, rubbing with a spreader rod, decanting off the liquid, and passing the resulting cloudy sporangia suspension through 50–70 μm nylon mesh or a cell strainer placed over a beaker. Add an equal volume of ALBA broth, mix gently, and count an aliquot using a hemocytometer. Dilute to 2.5×10^5 sporangia per mL by adding equal amounts of water and ALBA. Add about 150 mL of the sporangia mix to individual 1 L sterile flasks; avoid the temptation to put too much volume per flask as this may inhibit germination. Normally the ten 150-mm plates yield enough sporangia for four flasks.
3. Incubate the flasks at 18 °C without shaking for 24–36 h to allow the sporangia to germinate. The culture is normally ready to harvest when a thin mat of hyphae detaches from the bottom of the flask when swirled (*see* **Note 10**).
4. Collect the mat by pouring the culture through the nylon mesh. Any hyphae sticking to the flask can be scraped off using the tip of a pipette. The flow-through will contain mostly ungerminated sporangia; we recommend that this be incubated for another 24–36 h to yield tissue for a subsequent round of protoplasting, as this raises the efficiency of the procedure. Lift the hyphae from the nylon mesh using forceps, place in a 50-mL conical, and take note of the volume.
5. For every mL of loosely packed hyphae, prepare 3 mL of 10 mg/mL β-glucanase and 5 mg/mL cellulase in FPB. Typically, about 25 mL will be required. Sterilize using a syringe filter, placing the flow-through into a 50 mL conical tube.
6. Wash the hyphae from **step 4** once by adding 40 mL FPB, inverting, and passing through the nylon mesh or cell strainer. Lift out the washed mycelia using forceps and add to the tube with the protoplast enzymes (*see* **Note 11**).
7. Place the tube on its side on a rotary shaker. Shake at 50 rpm until protoplasting is about 90% complete, which should take about 25 min.
8. Filter through the cell strainer or nylon mesh, saving the flow-through in a beaker or 50 mL conical tube. Save a small amount to count the concentration. Pellet the protoplasts in a conical

tube by spinning at 700× *g* for 4 min in a swinging bucket rotor. While spinning, use a hemocytometer to count the starting protoplasts; anticipate losing about half in the following wash steps.

9. Carefully pour off the supernatant; the pellet may be soft. Use a pipette to remove any remaining drops of enzyme mix and resuspend the pellet in 30 mL FPB by stirring gently with a pipette. At this stage, thoroughly resuspending the pellet is not required and pipetting up and down is not recommended. Spin again at 700× *g* for 4 min.
10. Although DNA can be added directly to the protoplasts, we normally have better results by complexing the DNA with Lipofectin (*see* **Note 3**). It takes about 10 min for liposomes to form, so while the protoplasts are spinning, place 30 μg of plasmid DNA into a sterile polystyrene tube for each anticipated reaction (2×10^7 protoplasts for each 0.7 mL transformation); avoid polypropylene since this binds the liposomes. Add H_2O to a total of 40 μL, add 60 μL of Lipofectin, and gently pipette up and down or flick to mix. The mixture should turn cloudy. Reactions can be scaled up by increasing all components here and in subsequent steps in proportion.
11. Discard the supernatant from **step 9**, add 30 mL of 1:1 FPB: MT, and mix gently by stirring with a pipette; it is not necessary to thoroughly resuspend the pellet at this time. Spin at 700× *g* for 4 min.
12. Discard supernatant and add 30 mL MT supplemented with 10 mM $CaCl_2$. After thoroughly but gently resuspending the pellet, save a drop for counting on a hemocytometer, and spin the rest at 700× *g* for 4 min.
13. Discard the supernatant and resuspend pellet in enough MT supplemented with 10 mM $CaCl_2$ to have 2×10^7 of protoplasts per 0.7 mL reaction based on the count from Step 12. There should be enough protoplasts for about five reactions.
14. Add 0.7 mL of 50% PEG-MW 3350 containing 25 mM $CaCl_2$ and 10 mM Tris pH 7.5 to the protoplast-DNA mixture. Rotate or gently invert to mix, and incubate for 3–4 min.
15. Add clarified rye broth containing 1 M mannitol using the following stepwise procedure. First, add 2 mL of rye-mannitol to the protoplasts, and mix by inversion. After 30 s, add 4 mL of rye-mannitol, divide into two 50 mL tubes prefilled with 20 mL rye-mannitol, and invert. Lay the tube on its side and incubate at 18 to 25 °C for 18 to 30 h (*see* **Note 12**).
16. Check a drop from each tube for signs of regeneration, which is indicated by hyphae extending from the now-walled protoplasts. Using the entire tube, spin down at 700× *g* for 5 min

and decant the supernatant, leaving about 1 mL of free fluid in the tube. Gently resuspend the pellet in this residual liquid and spread 0.2 mL per 100-mm plate of rye-sucrose agar containing 8 μg/mL G-418. Incubate at 18 to 25 °C for 7 d, turning the plate upside down when the liquid is absorbed.

17. Start checking plates for colonies at day 6. These may continue to appear for up to 10 days while those appearing later are usually escapes resulting from G-418 breakdown. A successful transformation may yield 20–40 colonies per 2×10^7 of protoplasts. Transfer the colonies to fresh plates containing 8 μg/mL G-418 while they are still well-separated. If subsequent studies indicate that editing has occurred, G-418 does not need to be included in future cultures.

3.4 Extracting DNA from *Phytophthora* Transformants (See Note 13)

1. While colonies can be checked for editing on the original selection plates, we usually test them after being transferred to fresh plates.
2. Scrape a 1 cm^2 patch of hyphae from each culture, immerse in 0.3 mL extraction buffer in a 1.5 mL microcentrifuge tube, and place in a boiling water bath for 5 min.
3. Add 0.3 mL 1:1 phenol:chloroform. Vortex at high speed for 5 min, then spin for 5 min at 12,000× *g* or higher.
4. Pipette off 0.25 mL of the aqueous phase, place into a new 1.5 mL tube, and add 0.75 vol (0.187 mL) of 100% isopropanol. Mix well and spin for 10 min at 12,000× *g* or higher.
5. Wash the pellet by decanting the supernatant, adding 0.2 mL of 70% ethanol, spinning for 2 min, removing the supernatant, and air-drying the pellet for about 15 min.
6. Dissolve the DNA pellet in 30 μL of Low TE. This can be facilitated by heating at 65 °C for 5 min.

3.5 Analysis of Target Region

1. Obtain PCR primers flanking the gRNA target(s). We recommend using primers between 19 and 22 nt in size that yield an amplicon smaller than 800 nt, spaced at least 75 nt from each target.
2. For PCR, add 1 μL of the genomic DNA from Subheading 3.4 to a 25 μL reaction containing 1x PCR buffer, 1 mM $MgCl_2$, 200 μM dNTPs, 1 unit of *Taq* polymerase, and 0.2 μM of each primer. We normally perform 35-cycle reactions that include a 30 s denaturation step at 95 °C, 30 s of annealing at 50–60 °C depending on the primer, and 1 min of extension at 72 °C.
3. Visualize the products on a 1% agarose gel with a DNA stain. A single band sized differently than the unedited target will indicate biallelic editing resulting from NHEJ or monoallelic editing followed by gene conversion (Fig. 2a). A mixture of normal

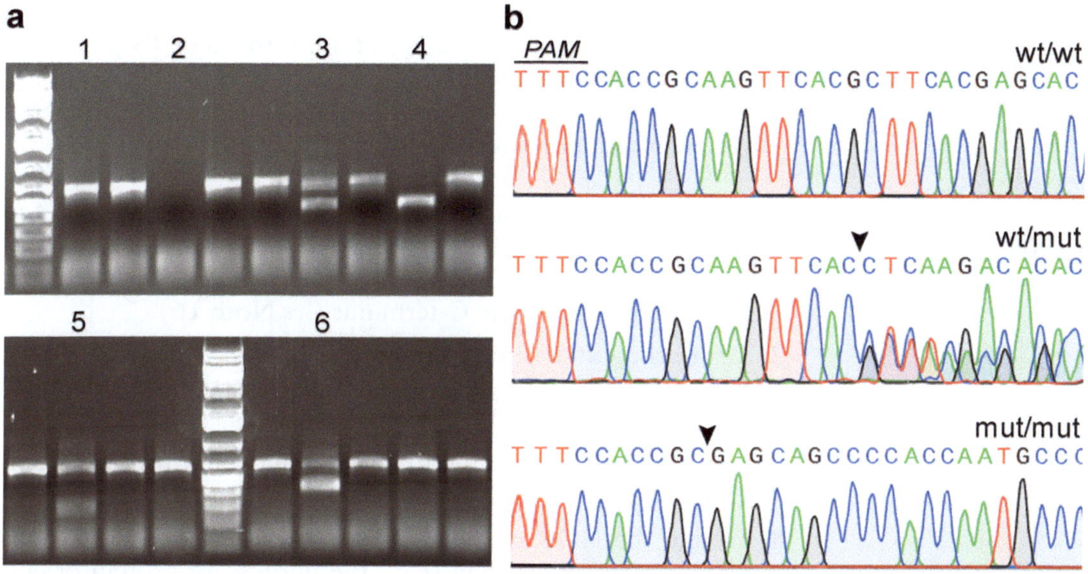

Fig. 2 Possible outcomes of editing. (**a**) PCR of target region in transformants. The lane labelled 1 is a representative unedited strain. Lane 2 is a strain in which a large deletion removed one or both primer sites; this would need to be re-screened with other primers to determine the precise nature of the editing event. Lane 3 is a heterozygote in which a monoallelic deletion resulted in two bands of equal intensity representing the unedited and edited alleles, respectively. Lane 4 is a biallelic edited strain; this possibly resulted from editing of one allele followed by gene conversion. Lane 5 is a likely heterokaryon containing both edited and unedited nuclei. The dominance of the wild-type band suggests heterokaryosis rather than heterozygosity. Lane 6, another possible heterokaryon in which the edited allele is stronger; further experimentation would be required to determine if this results from heterokaryosis or preferential amplification of the smaller band. (**b**) Sequencing chromatograms of transformants, starting from the PAM. The arrowheads indicate the 5′ junctions of the edited regions. Shown are an unedited strain (wt/wt), one in which a single allele was edited based on the presence of double peaks (wt/mut), and a strain containing two identical edited alleles (mut/mut)

and altered bands signals monoallelic editing. If the band ratio diverges much from 1:1, editing may have occurred after the first nuclear division, resulting in a heterokaryon. An absence of a band means either a large deletion or problems with DNA extraction or PCR (*see* **Note 14**).

4. Submit samples for Sanger sequencing (*see* **Notes 6** and **15**). This is done since small indels may not be evident from gel analysis, plus the results will allow the indel to be sized precisely to check for a frameshift that may block protein function. Potential outcomes are illustrated in Fig. 2b.

5. Heterokaryons containing a mixture of edited and unedited nuclei can often be converted to biallelic edited strains by single-zoospore (single nuclear) purification. This is also recommended for strains that will be used for experimental analyses such as phenotyping. This is achieved by isolating sporangia as in Subheading 3.3, **step 1** incubating them for

2 h at 10 °C to stimulate the release of zoospores, separating zoospores from sporangia by filtration through 15 μm mesh, and plating the zoospores at low density on selective media. Examination using a dissecting microscope can help identify incipient zoospore-derived colonies.

3.6 Protein Tagging by HDR

1. Plan the overall strategy by identifying the desired tag and whether the fusion should be N-terminal, C-terminal, or internal to the protein of interest. Illustrated in Fig. 3 is a strategy for adding a tag to the C-terminus (*see* **Note 16**).
2. Using the criteria in Subheading 3.1, design a gRNA that is preferably within 50 bp of the desired insertion site (*see* **Note 17**).
3. Once gRNA(s) are chosen, they should be synthesized and cloned into the Cas12a plasmid as in Subheading 3.2.
4. Design the HDR repair template, i.e., the donor DNA. This needs to include the tag (e.g., the fluorescent protein or epitope) flanked by homology arms that match the chromosomal target. We have had success with homology arms as small as 100 nt and as large as 500 nt. Since the optimal length of the homology arm may vary with the chromosome target and the distance between the Cas12a cut site and the desired location of recombination, we normally take a conservative approach of using arms between 500 and 1000 nt. The use of large arms was recommended in a prior study in *P. sojae* [2].

A short linker (~20 amino acids) should be provided to connect the tag with the protein of interest. The PAM or gRNA sequences within the donor should be altered to prevent re-cleavage by Cas12a, using silent point mutations if within the coding sequence (*see* **Note 18**).

5. Prepare the donor template with the following structure: left homology arm–tag–right homology arm. This is illustrated in Fig. 3 for a C-terminal mScarlet tag. The donor DNA is usually constructed in a general plasmid such as pBluescript using synthesized DNA as in Subheading 3.2, or by a ligation-independent long homology-based cloning method [27]. For HDR, we normally linearize the donor using one of the following three strategies. Option 1 is to linearize the plasmid prior to transformation by digestion with a restriction enzyme flanking one or both homology arms. Option 2 is to generate the linear DNA by PCR using primers binding the ends of the two homology arms. Option 3 is to construct a donor plasmid in which gRNA-PAM recognition sequences (the same as in **step 2**) are on the 5′ and 3′ ends of the homology arms, as shown in Fig. 3c. The latter approach relies on cleavage of the donor DNA in vivo during transformation (*see* **Note 19**).

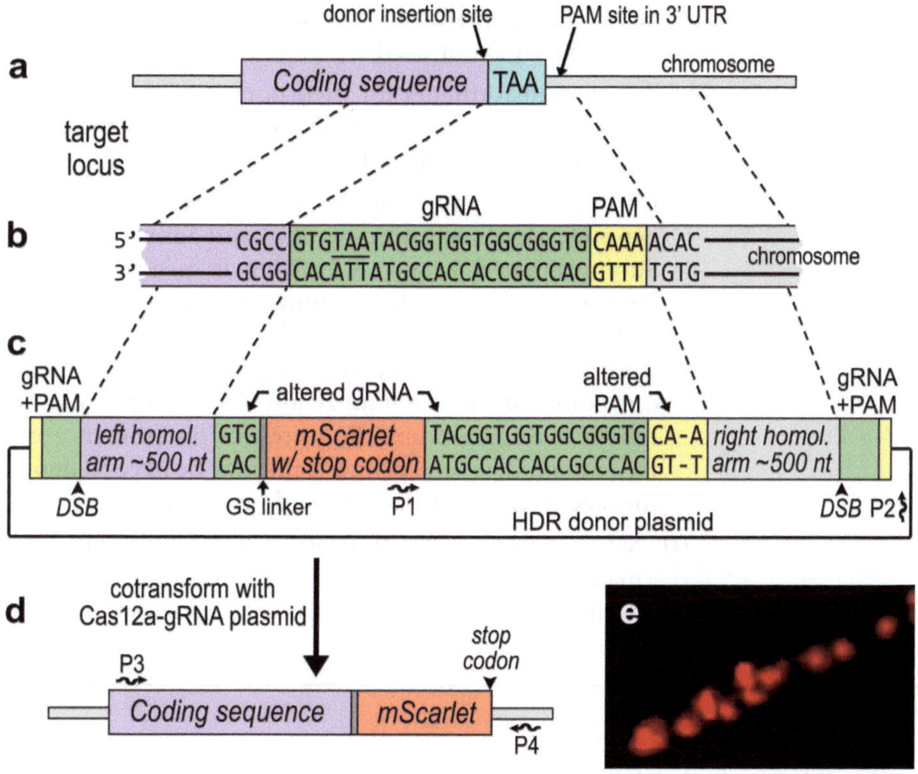

Fig. 3 Outline of C-terminal HDR tagging using mScarlet. (**a**) Schematic of a chromosomal target, indicating the desired insertion site 5′ of the stop codon and a PAM in the 3′ UTR. Having the gRNA within 50 nt of the insertion site is recommended to maximize HDR. (**b**) Closeup of the targeted region showing representative PAM and gRNA sequences. In this example, the stop codon is within the gRNA site (underlined), but this is not required for tagging since HDR results mostly from recombination within the homology arms, distal from the gRNA. (**c**) Diagram of the donor plasmid. This contains at least 500 nt of left and right homology arms located 5′ and 3′ of the insertion site, respectively; PAM and gRNA sequences added outside the homology arms to cause double-strand breaks (DSB) during transformation; a linker fused upstream of mScarlet (GS); a stop codon at the 3′ end of mScarlet; and an altered PAM to prevent continued editing by Cas12a after HDR. The placement of mScarlet within the gRNA site will also block continued cleavage, although altering both the PAM and gRNA regions is unnecessary. P1 and P2 represent PCR primers used to test for plasmid integration at random sites; HDR should result in no amplification since P2 binds to plasmid DNA outside the homology arm. (**d**) Structure of the chromosomal locus structure after HDR. P3 and P4 represent PCR primers used to help confirm in-frame incorporation of mScarlet and test if one or both alleles were tagged. These primers should reside distal to the homology arms. (**e**) Hyphal tip from a transformant in which a transcription factor was tagged with mScarlet. The protein localized to nuclei as expected

6. After obtaining sufficient DNA for transformation, co-transform *P. infestans* with the Cas12a-gRNA plasmid and the HDR donor plasmid (or linear DNA) in a 1:2.5 molar ratio. This is achieved as in Subheading 3.3 using a total of 35 to 50 μg DNA. A molar excess of donor template should raise the number of HDR events, although this may come at the cost of more off-target integrations.

7. Select transformants on rye containing G-418; a marker for the donor DNA is not used.
8. Screen transformants for incorporation of the tag. If using a fluorescent tag, we normally first test for its expression by microscopy (*see* **Note 20**).
9. Confirm the nature of the integration event by extracting genomic DNA, followed by PCR and Sanger sequencing as in Subheadings 3.4 and 3.5. This is needed to distinguish between precisely edited clones and off-target integrations, and between mono- and biallelic editing. For downstream studies, we recommend selecting transformants bearing only one edited allele to minimize any effects of tagging on protein function. Suggested binding sites for the primers are shown in Fig. 3c and d; several primer pairs may be needed depending on the size of the genes and homology arms.
10. Single-zoospore purification of successfully tagged transformants is recommended to isolate a pure strain containing the desired heterozygous or homozygous knock-ins.

4 Notes

1. pSTUC1 is a derivative of Cas12a, which includes amino acid changes that allow editing to function better over a broad temperature range [28]. We recommend using this variant for species like *P. infestans* that grow at cooler temperatures, e.g., 18–22 °C. We have not tested whether this variant is superior for species of *Phytophthora* having higher optimal growth temperatures, e.g., 28 °C or above.
2. Some published editing protocols using Cas9 include a step for phosphorylating the oligonucleotides, but this is not required unless the vector has been dephosphorylated to block self-ligation. The latter scenario should not occur with our vectors since their two *Bsa*I sites do not generate cohesive ends.
3. We normally use a *Trichoderma* cellulase purchased from MilliporeSigma (catalog number C8546), but most other cellulases work equally well. We have used several β-glucanases with success but found that the most effective and economical choices are products developed for clarifying wines, such as Vinoflow NCE (Novozymes) and Extralyse (Laffort).
4. We have found that Lipofectin increases the number of transformants by two to threefold, but this can be omitted.
5. The drug concentration may need to be varied depending on the strain, species, or lot of chemical. We suggest performing a preliminary experiment to identify the minimal concentration that blocks growth. Other types of media may also be suitable

for species besides *P. infestans*. We have also constructed editing vectors using hygromycin as the selectable marker, as G-418 may not be suitable for all oomycetes.

6. PCR reactions must be purified prior to DNA sequencing to remove primers and nucleotide triphosphates. This can be done using a PCR clean-up column or by treatment with exonuclease I and (shrimp) alkaline phosphatase. We find the latter to be more convenient. Both enzymes are available from New England Biolabs, or packaged together in the ExoSAP-IT kit from ThermoFisher or the Exo-Alp PCR Cleanup Mix kit from AdvancedSeq.
7. The gRNA and target should match perfectly within the seed region, which for Cas12a is the first 5–6 bases after the PAM. Up to two mismatches elsewhere often do not severely impair editing [29, 30]. Nevertheless, we aim for monomorphic sites to maximize editing efficiency. Monomorphism within the binding sites for PCR primers used to test for editing is also important to ensure that both alleles will be amplified. If the gRNAs were designed using a genome from a strain other than the one being transformed, the occurrence of inter-strain polymorphisms should also be considered. Ideally, such issues can be addressed by examining a phased genome assembly of the strain that will be edited. Alternatively, one can search unassembled DNA sequence reads from Illumina or other technologies; such data can also be used to test for diploidy considering that knocking out genes in polyploids may be more challenging [8]. Data about polymorphisms might be obtained by sequencing PCR products from the genomic target, but this assumes that the primers would bind both alleles. RNA reads may also be useful, although we have observed cases where the alternate alleles of a gene were transcribed at very dissonant levels [3].
8. To ensure maximum yield, we recommend streaking out a fresh plate of a bacterial colony containing the vector. Pick a colony and grow in 2 mL of LB or TB both containing 100 μg/mL carbenicillin to exponential phase (4–8 h) at 37 °C with 250 rpm; this antibiotic persists longer in media than ampicillin. Inoculate 0.1% volume (250 μL) into 250 mL of TB containing 100 μg/mL carbenicillin in a 1 L flask. Incubate at 37 ° C and 250 rpm for no longer than 12 h prior to DNA extraction; the total yield is normally 0.3 to 1.2 mg.
9. Several other techniques exist for transformation, including zoospore electroporation and *Agrobacterium*-mediated transformation [11, 12]. We usually employ the protoplast method for editing since it can yield the most transformants, but good evidence that the resulting transformants exhibit higher rates of editing is lacking.

10. Obtaining young tissue for protoplasting is important. Older cultures will digest poorly and yield many anucleate protoplasts, which fruitlessly soak up DNA. Our protocol is optimized for *P. infestans* isolate 1306 to achieve more synchronous germination of sporangia, yielding more young hyphae and hence, more transformants. The timepoints may need to be altered for other strains, and in some cases, it may be useful to test other approaches for obtaining young hyphae such as by germinating zoospores in media.
11. It is important to move quickly through the succeeding steps to minimize the time that the tissue and protoplasts are not in media. To help accomplish this, in the next step we collect the protoplasts before the hyphae is 100% digested.
12. Although our previously published transformation procedures describe incubating protoplasts at 18 °C following DNA treatment, for editing we recommend regenerating the protoplasts for one day at 25 °C prior to spreading on agar media. This is above the optimal growth temperature for *P. infestans,* but increases the frequency of editing [28]. After plating on agar media, cultures can be shifted to 18–20 °C since most editing events appear to occur within a day after treating the protoplasts with DNA. We recommend spreading the regenerated tissue on agar media after no more than 24 h of growth in broth, since longer incubation times will result in matted hyphae that are difficult to distribute on the plates. We also recommend including an antifungal and antibacterial in the regeneration broth, such as 10,000 units/mL of nystatin and 50 μg/ml penicillin G.
13. We have found that this method yields more reliable results in PCR than DNA from column-based kits.
14. The editing rate will vary depending on the gene, gRNA, and transformation reaction. While we have observed NHEJ in up to 75% of transformants in some experiments, a frequency of 15 to 20% is more typical. Cases in which no editing or only non-frameshift mutations are observed may signal that the target is an essential gene.
15. DNA for sequencing can be isolated directly from the PCR sample. If nonspecific bands are present, we advise gel purification or, better yet, repeating the amplification with more specific primers. Otherwise, the minor bands may yield peaks in chromatograms that would erroneously suggest editing.
16. The location for the tag should be tailored to the protein. For example, a tag should not be placed upstream of a gene for a secreted protein as this may interfere with localization. If a fluorescent reporter is used, choose one that is photostable, bright, and imageable with equipment available to your lab.

When expressed as a fusion protein, monomeric tags such as mScarlet have less tendency to induce toxicity than proteins such as GFP which may dimerize [31, 32].

17. A 50 nt window is suggested since the frequency of HDR may decline with increasing distance from the desired insertion site [33]. Note that the double-strand break generated by Cas12a serves only as the site to initiate repair. The final exchange points between chromosome and donor will be within the homology arms, potentially distant from the double-strand break. Nevertheless, placing the gRNA in a non-coding region is preferred to avoid altering the protein's sequence outside of the tag.
18. Linkers are added to minimize the risk of sterically hindering protein function. Their optimal sequence depends on the nature of the tag and its desired application [34]. Nevertheless, most linkers consist of amino acids such as glycine and serine that provide flexibility [35]; we had success with $(GGGGS)_2$. Besides adding a linker, remember to design the donor to eliminate unwanted stop codons.
19. Although a linear donor is not essential for HDR, most studies have reported higher efficiencies with linear versus circular (plasmid) donors [36]. We have usually generated a linear donor using Option 3 due to reports that this improved HDR in other systems [37]. The theory is that the simultaneous activation of the target site and in vivo linearization of the donor construct synchronizes the demand and supply of homologous sequences, thus increasing HDR. However, data comparing HDR efficiencies in *Phytophthora* using circular, single-cut linearized, or double-cut linearized templates are not yet available.
20. Alternatively, Western blot analysis can be performed on cell lysates using antibodies against the fluorescent reporter or epitope tag. This will also confirm if the fusion protein has the expected size.

Acknowledgments

This work has been supported by grants to H.S.J. from the National Institute of Food and Agriculture of the US Department of Agriculture, and the National Science Foundation of the United States.

References

1. Anzalone AV, Koblan LW, Liu DR (2020) Genome editing with CRISPR-Cas nucleases, base editors, transposases and prime editors. Nat Biotechnol 38:824–844
2. Fang Y, Tyler BM (2016) Efficient disruption and replacement of an effector gene in the oomycete *Phytophthora sojae* using CRISPR/Cas9. Mol Plant Pathol 17:127–139
3. Ah-Fong A, Boyd AM, Matson MEH, Judelson HS (2021) A Cas12a-based gene editing system for *Phytophthora infestans* reveals monoallelic expression of an elicitor. Mol Plant Pathol 22:737–752
4. Gumtow R, Wu DL, Uchida J, Tian MY (2018) A *Phytophthora palmivora* extracellular cystatin-like protease inhibitor targets papain to contribute to virulence on papaya. Mol Plant-Microbe Interact 31:363–373
5. Chen XR, Zhang Y, Li HY, Zhang ZH, Sheng GL, Li YP et al (2019) The RXLR effector PcAvh1 is required for full virulence of *Phytophthora capsici*. Mol Plant-Microbe Interact 32:986–1000
6. Erwin DC, Ribeiro OK (1996) *Phytophthora* diseases worldwide. APS Press St, Paul, Minn
7. Fry WE (2020) *Phytophthora infestans:* the itinerant invader; "late blight": the persistent disease. Phytoparasitica 48:87–94
8. Matson MEH, Liang Q, Lonardi S, Judelson HS (2022) Karyotype variation, spontaneous genome rearrangements affecting chemical insensitivity, and expression level polymorphisms in the plant pathogen *Phytophthora infestans* revealed using its first chromosome-scale assembly. PLoS Pathog 18:e1010869
9. Ah-Fong AM, Kim KS, Judelson HS (2017) RNA-seq of life stages of the oomycete *Phytophthora infestans* reveals dynamic changes in metabolic, signal transduction, and pathogenesis genes and a major role for calcium signaling in development. BMC Genomics 18:198
10. Ah-Fong AM, Bormann-Chung CA, Judelson HS (2008) Optimization of transgene-mediated silencing in *Phytophthora infestans* and its association with small-interfering RNAs. Fungal Genet Biol 45:1197–1205
11. Wu DL, Navet N, Liu YC, Uchida J, Tian MY (2016) Establishment of a simple and efficient *agrobacterium*-mediated transformation system for *Phytophthora palmivora*. BMC Microbiol 16:204
12. Ah-Fong AMV, Kagda M, Judelson HS (2018) Illuminating *Phytophthora* biology with fluorescent protein tags. In: Plant pathogenic fungi and oomycetes: methods and protocols, Methods in Molecular Biology, vol 1848. Humana Press Inc, Totowa, NJ, pp 119–129
13. Ah-Fong AM, Judelson HS (2011) Vectors for fluorescent protein tagging in *Phytophthora*: tools for functional genomics and cell biology. Fungal Biol 115:882–890
14. Yeh CD, Richardson CD, Corn JE (2019) Advances in genome editing through control of DNA repair pathways. Nat Cell Biol 21: 1468–1478
15. Shin S, Kim SH, Shin SW, Grav LM, Pedersen LE, Lee JS et al (2020) Comprehensive analysis of genomic safe harbors as target sites for stable expression of the heterologous gene in HEK293 cells. ACS Synth Biol 9:1263–1269
16. Bukhari H, Muller T (2019) Endogenous fluorescence tagging by CRISPR. Trends Cell Biol 29:912–928
17. Saeed S, Usman B, Shim SH, Khan SU, Nizamuddin S, Saeed S et al (2022) CRISPR/Cas-mediated editing of cis-regulatory elements for crop improvement. Plant Sci 324:111435
18. Zetsche B, Gootenberg JS, Abudayyeh OO, Slaymaker IM, Makarova KS, Essletzbichler P et al (2015) Cpf1 is a single-RNA-guided endonuclease of a class 2 CRISPR-Cas system. Cell 163:759–771
19. Moreno-Mateos MA, Fernandez JP, Rouet R, Vejnar CE, Lane MA, Mis E et al (2017) CRISPR-Cpf1 mediates efficient homology-directed repair and temperature-controlled genome editing. Nature Comm 8:2024
20. Fonfara I, Richter H, Bratovic M, Le Rhun A, Charpentier E (2016) The CRISPR-associated DNA-cleaving enzyme Cpf1 also processes precursor CRISPR RNA. Nature 532:517–521
21. Zetsche B, Heidenreich M, Mohanraju P, Fedorova I, Kneppers J, DeGennaro EM et al (2017) Multiplex gene editing by CRISPR-Cpf1 using a single crRNA array. Nat Biotechnol 35:31–34
22. Kim D, Kim J, Hur JK, Been KW, Yoon SH, Kim JS (2016) Genome-wide analysis reveals specificities of Cpf1 endonucleases in human cells. Nat Biotechnol 34:863–868
23. Wang Q, Coleman JJ (2019) Progress and challenges: development and implementation of CRISPR/Cas9 technology in filamentous fungi. Comput Struct Biotec 17:761–769
24. Peng D, Tarleton R (2015) EuPaGDT: a web tool tailored to design CRISPR guide RNAs for eukaryotic pathogens. Microb Genom 1: e000033

25. Kim HK, Min S, Song M, Jung S, Choi JW, Kim Y et al (2018) Deep learning improves prediction of CRISPR-Cpf1 guide RNA activity. Nat Biotechnol 36:239–241
26. Jensen KT, Floe L, Petersen TS, Huang JR, Xu FP, Bolund L et al (2017) Chromatin accessibility and guide sequence secondary structure affect CRISPR-Cas9 gene editing efficiency. FEBS Lett 591:1892–1901
27. Singh V (2020) Recent advances, challenges, and opportunities in synthetic genomes. In: Singh V (ed) Advances in synthetic biology. Springer, Singapore, pp 221–235
28. Mendoza CS, Findlay AC, Judelson HS (2023) An LbCas12a variant and elevated incubation temperatures enhance the rate of gene editing in the oomycete *Phytophthora infestans*. Mol Plant-Microbe Interact 36:677. https://doi.org/10.1094/MPMI-05-23-0072-SC
29. Kleinstiver BP, Tsai SQ, Prew MS, Nguyen NT, Welch MM, Lopez JM et al (2016) Genome-wide specificities of CRISPR-Cas Cpf1 nucleases in human cells. Nat Biotechnol 34:869–874
30. Murugan K, Seetharam AS, Severin AJ, Sashital DG (2020) CRISPR-Cas12a has widespread off-target and dsDNA-nicking effects. J Biol Chem 295:5538–5553
31. Bindels DS, Haarbosch L, van Weeren L, Postma M, Wieser KE, Mastop M et al (2017) mScarlet: a bright monomeric red fluorescent protein for cellular imaging. Nat Methods 14:53–56
32. Kleeman B, Olsson A, Newkold T, Kofron M, DeLay M, Hildeman D et al (2018) A guide to choosing fluorescent protein combinations for flow cytometric analysis based on spectral overlap. Cytom Part A 93a:556–562
33. Elliott B, Richardson C, Winderbaum J, Nickoloff JA, Jasin M (1998) Gene conversion tracts from double-strand break repair in mammalian cells. Mol Cell Biol 18:93–101
34. Patel DK, Menon DV, Patel DH, Dave G (2022) Linkers: a synergistic way for the synthesis of chimeric proteins. Protein Expres Purif 191:106012
35. Snapp E (2005) Design and use of fluorescent fusion proteins in cell biology. Curr Protoc Cell Biol Chapter 21:21.4.1–21.4.13
36. Song F, Stieger K (2017) Optimizing the DNA donor template for homology-directed repair of double-strand breaks. Mol Ther Nucleic Acids 7:53–60
37. Wolter F, Puchta H (2019) In planta gene targeting can be enhanced by the use of CRISPR/Cas12a. Plant J 100:1083–1094
38. Strohkendl I, Saifuddin FA, Rybarski JR, Finkelstein IJ, Russell R (2018) Kinetic basis for DNA target specificity of CRISPR-Cas12a. Mol Cell 71:816–824

Chapter 5

CRISPR/Cas9-Mediated Gene-Knockout and In Situ Complementation System for *Phytophthora sojae*

Min Qiu, Nan Chen, and Yuanchao Wang

Abstract

The establishment of reliable and efficient systems for genome editing in *Phytophthora* is very important for studying gene functions. Here, step-by-step methods for CRISPR/Cas9-based gene knockout and in situ complementation for *Phytophthora sojae* are presented. These steps include the sgRNA design, Cas9-sgRNA plasmid construction, homologous replacement, complementation vector construction, *P. sojae* transformation, and detection of mutations for both gene knockout and in situ complementation. These methods may also potentially be adapted for other *Phytophthora* species.

Key words CRISPR/Cas9, Gene knockout, Gene complementation, Gene editing, *Phytophthora sojae*

1 Introduction

The low probability of homologous recombination during the transformation of *Phytophthora* species makes it difficult to knock out genes. Thus, to enable efficient and accurate gene editing, CRISPR/Cas9-based technology was developed for *P. sojae* by Fang and Tyler, which involves the use of Cas9 and single guide RNA (sgRNA) [1]. In this method, a 20-nucleotide single guide RNA mediates the targeting of Cas9 to specific sequences based on the double-stranded complementarity principle. The Cas9 nuclease cleaves specific DNA sites and creates double-strand break gaps. In the presence of a homologous template (donor DNA), cells use the homologous template for repair, resulting in knockout or disruption and replacement of the target gene with the donor DNA.

To confirm that it is the knockout of the targeted gene that leads to a specific phenotype, an in situ complementation assay should be performed to complete the functional gene analysis. In order to avoid Cas9 targeting a newly transformed complementing gene, a functioning copy of the target gene, but with mutated

Monica L. Gerth and Rosie E. Bradshaw (eds.), *Phytophthora: Methods and Protocols*, Methods in Molecular Biology, vol. 2892, https://doi.org/10.1007/978-1-0716-4330-3_5,

sgRNA sites, is used as the donor DNA in the complementation system [2]. This mutated donor DNA vector is co-transformed into the *P. sojae* knockout line together with Cas9 and two new RNA templates targeting the substitution gene. In this way, the original knockout gene is substituted for the original complementation gene in situ, resulting in the reintroduction of the target gene at the original site.

Due to recent advances in *Phytophthora* studies, a CRISPR/Cas9-mediated knockout and in situ complementation system has been successfully established and implemented. This system will greatly facilitate the study of gene function in *Phytophthora*. This chapter provides the fundamental information needed to execute the gene knockout and in situ complementation in *P. sojae* and can be adapted for use with other *Phytophthora* species.

2 Materials

2.1 sgRNA Design

1. sgRNA design software: e.g., EuPaGDT (http://grna.ctegd.uga.edu/).
2. RNA secondary structure prediction tool, e.g., (http://rna.urmc.rochester.edu/RNAstructureWeb/Servers/Predict1/Predict1.html).

2.2 Construction of CRISPR/Cas9 Vector

1. Regular Taq DNA polymerase.
2. Plasmid: pYF515 (constructed by Fang and Tyler, 2017) (Fig. 1).
3. BsaI restriction enzyme and suitable 10× Buffer.

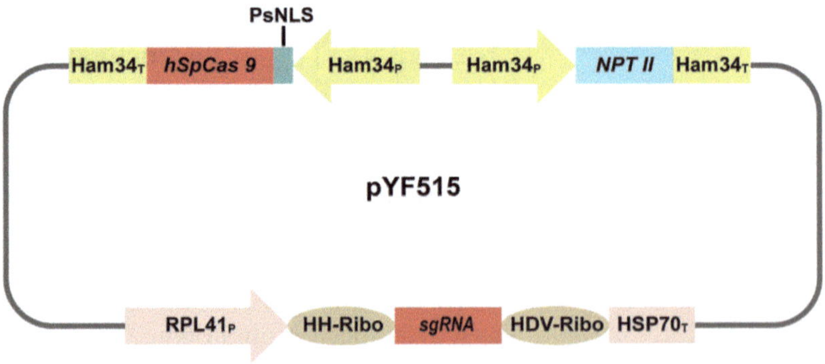

Fig. 1 The “all in one” plasmid pYF515 contains both the Cas9 and the sgRNA expression units. Cas9 expression, the nuclear localization sequence (*PsNLS*), and the selection marker *NPT II* are driven by the *Ham34* promoter. Transcription of single guide RNA (sgRNA) (including the flanking ribozymes: hammerhead (HH) ribozyme and hepatitis delta virus (HDV) ribozyme) is driven by the *RPL41* promoter [1]

4. NheI restriction enzyme and 10× Buffer.
5. Water bath (42 °C).
6. PCR purification kit.
7. T4 DNA Ligase and 10× T4 DNA Ligase Buffer.
8. *Escherichia coli* JM109 (or similar cloning strain) competent cells.
9. Luria-Bertani medium: 10 g Bacto Tryptone, 5 g Yeast Extract, 10 g NaCl and make up to 1 L with distilled water. Add 1.5% agar (w/v, for solid medium).
10. Plasmid extraction Mini Kit.
11. Ampicillin solution: 50 mg/mL in water.

2.3 Construction of Replacement and Complementation Gene Vectors

1. Gel DNA Extraction Mini Kit.
2. Recombinational Cloning Kit.
3. Plasmid: pBluescript SK II+ (pBS-SK II+).
4. Items 8–11 from Subheading 2, **item 2.**

2.4 PEG-Mediated Stable Transformation of P. sojae Protoplasts

1. *Phytophthora sojae* strain P6497 [3].
2. Clarified pea broth: Mix 120 g of frozen peas and 700 mL of distilled water together. Autoclave and collect filtrate by fine-mesh gauze filtration.
3. Vitamin solution: 0.012 g L-inositol, 0.06 g nicotinic acid, 0.18 g pyridoxine–HCl, 0.015 g riboflavin, and 0.38 g thiamine-HCl. Add 10 μL of 0.02 g/mL biotin, 10 μL of 0.02 g/mL folic acid and make up to 300 mL with distilled water. Filter-sterilize and then store (up to 1 year) at 4 °C.
4. Trace elements: 0.215 g $FeC_6H_5O_7{\cdot}3H_2O$, 0.15 g $ZnSO_4{\cdot}7H_2O$, 0.03 g $CuSO_4{\cdot}5H_2O$, 0.015 g $MgSO_4{\cdot}H_2O$, 0.01 g H_3BO_3, 0.007 g MoO_3. Make up to 400 mL with distilled water. Filter-sterilize and store (up to 1 year) at 4 °C.
5. Nutrient Pea Broth (NPB) or NPB-Agar: 1.0 g KH_2PO_4, 1.0 g K_2HPO_4, 3.0 g KNO_3, 0.5 g $MgSO_4$, 0.1 g $CaCl_2$, 2.0 g $CaCO_3$, 5.0 g D sorbitol, 5.0 g D-mannitol, 5.0 g glucose, 2.0 g yeast extract and add to 700 mL of clarified pea broth (see Subheading 2, **item 4** Part 2). Add 2.0 mL of vitamin solution and 2.0 mL of trace elements and make up to 1 L with distilled water. For NPB-agar add 10 g agar prior to autoclaving. Autoclave at 121 °C for 20 min and store at 4 °C.
6. Pea and Mannitol (PM) broth or PM-agar: 91.1 g mannitol, 1.32 g $CaCl_2{\cdot}2H_2O$, 2 g $CaCO_3$, and add to 700 mL pea broth supernatant (see Subheading 2, **item 4** Part 2). Make up to 1 L with water. For PM-agar, add 10 g bacto agar prior to autoclaving. Autoclave at 121 °C for 20 min and store at room temperature. Pre-cool to 4 °C before use.

7. 10% V8 broth or agar: Dissolve 1 g $CaCO_3$ in 100 mL of V8 vegetable juice (Campbell Soup Company), then add 900 mL of distilled water. For V8-agar, 15 g agar prior to autoclaving. Store at room temperature.
8. 0.5 M MES-KOH solution: Mix 4.88 g MES with 40 mL of distilled water and adjust pH to 5.7 with 1 M KOH. Filter sterilize and store at 4 °C.
9. Enzyme solution: Mix 10 mL of 0.8 M mannitol, 0.8 mL of 0.5 M KCl, 0.8 mL of 0.5 M MES, 0.4 mL of 0.5 M $CaCl_2$ with 0.15 g Lysing enzymes blend from *Trichoderma harzianum* (Sigma-Aldrich), 0.06 g Cellylysin cellulase and 8 mL of distilled water. Filter-sterilize into a sterile 50 mL centrifuge tube using a 0.45 μm syringe filter. The enzyme solution should be freshly prepared for each use.
10. 40% PEG solution: Mix 6 g PEG4000 and 3.75 mL of 0.8 M mannitol with 3 mL of distilled water and 3 mL of 0.5 M $CaCl_2$. Stir until completely dissolved. The PEG solution should be freshly prepared, filter-sterilized with a 0.45 μm sterile syringe filter into a 50-mL sterile beaker and kept at 4 °C or on ice when in use.
11. 0.8 M mannitol solution: Weigh 145.76 g Mannitol and add distilled water to 1 L. Autoclave at 121 °C for 20 min and store at room temperature. Pre-cool to 4 °C before use.
12. W5 solution: Weigh 0.093 g KCl, 4.6 g $CaCl_2{\cdot}2H_2O$, 2.25 g NaCl, 7.8 g glucose and add distilled water to 250 mL. Autoclave at 121 °C for 20 min and store at 4 °C.
13. MMg solution: Weigh 18.22 g mannitol, 0.76 g $MgCl_2{\cdot}6H_2O$, 2 mL of 0.5 M MES (pH = 5.7) and add distilled water to 250 mL. Autoclave at 121 °C for 20 min and store at 4 °C.
14. Ampicillin stock solution (50 mg/ml). Prepare in water and sterilize by filtration.
15. G418 stock solution (50 mg/ml). Prepare in water and sterilize by filtration.
16. 250-mL Erlenmeyer flasks. Sterilize by autoclaving.
17. Miracloth (sterilize by autoclaving).
18. Beakers (50 mL, 200 mL).
19. 50 mL centrifuge tubes.
20. 50 mL round-bottom centrifuge tubes.
21. Sterile syringe filters (0.45 μm, 25 mm diameter).
22. Rotating mixer.
23. Centrifuge with pre-cooling to 4 °C (e.g., Eppendorf, model: 5810R).

2.5 Screening and Identification of Knockout and Complementation Mutants

1. 2% CTAB solution: 80 mL of 1 M Tris-HCl, 32 mL of 0.5 M EDTA, 65.52 g NaCl, and 16 g CTAB. Make up to 800 mL with water. Heat the solution in a 65 °C water bath until dissolved, then adjust the pH to 8.0 with 1 M NaOH.
2. Chloroform.
3. Absolute ethanol.
4. Incubator-shaker (capable of reaching 37 °C).
5. Microcentrifuge (e.g., Eppendorf, model: 5425).
6. Plant Genomic DNA Extraction Kit (e.g., DNAsecure, Tiangen Biotech).
7. Taq polymerase (e.g., TaKaRa Taq), 10× PCR Buffer (Mg^{2+} plus), and dNTP Mixture.
8. DNA electrophoresis apparatus.

3 Methods

3.1 Design and Analysis of Candidate sgRNAs

1. Use a CRISPR guide design tool such as EuPaGDT (http://grna.ctegd.uga.edu/) to find suitable sgRNA sites in the target gene sequence, following the tool instructions. If using EuPaGDT, enter the gene ID in the "Job Name" column, select the "SpCas9" option in the "RNA guided nuclease selection" column, select *P. sojae* P6497 from Oomycetes in the "Genome (expand a category and choose one)" column, enter the gene sequence in the "Sequence" column, and click "get guide RNA" to obtain all the potential sgRNA targets in the target sequence.
2. Determine the specificity and perform off-target analysis of the sgRNAs.

Select sgRNAs with high efficiency (sgRNA score ≥ 0.5) and good specificity. Use BLASTN to determine the specificity of the sgRNA sequence in the *P. sojae* (or other relevant) genome. If non-target sequences in the genome perfectly match the final 15 nucleotides of the sgRNAs (including NGG), the sgRNA should be discarded. Subsequently, predict the secondary structure of the candidate sgRNAs using an RNA secondary structure prediction website such as: http://rna.urmc.rochester.edu/RNAstructureWeb/Servers/Predict1/Predict1.html sgRNAs with fewer than three consecutive self-complementary sequence pairings should be selected.

3.2 Construction of Gene Knockout CRISPR/Cas9 Vector

1. sgRNA synthesis-generic primer design with plasmid pYF515. The designed sgRNA needs to be incorporated into the "all in one" plasmid pYF515 which is used to express Cas9 as well as the sgRNA [4] (*see* **Note 1**). Once constructed, the sgRNA will

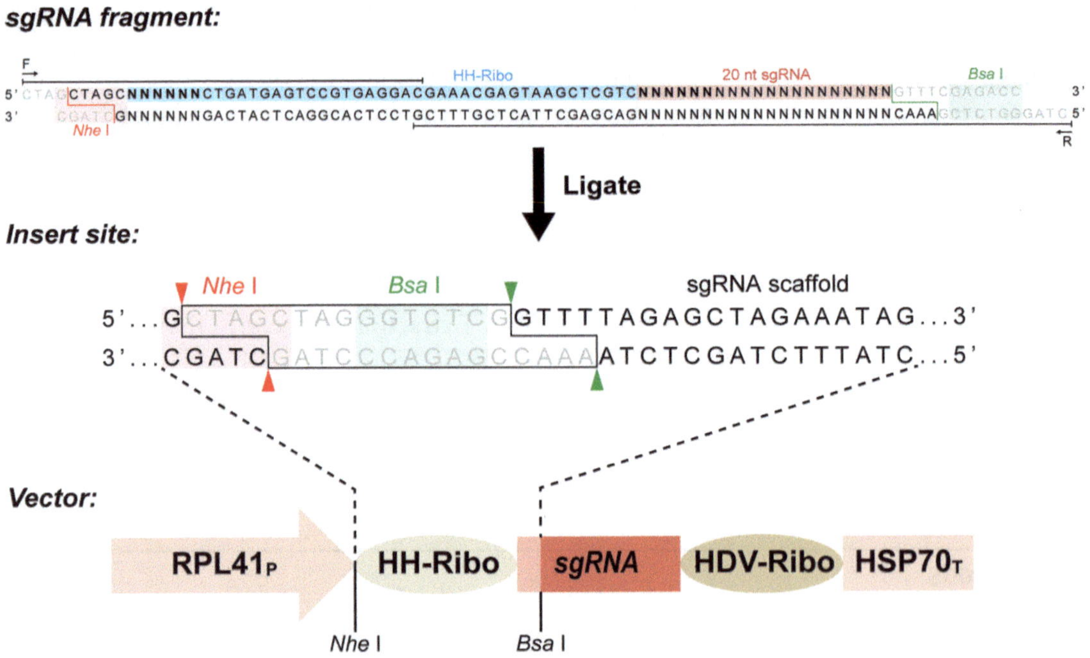

Fig. 2 Construction of gene knockout CRISPR/Cas9 vector. The target sgRNA fragment includes both the HH-Ribo sequence and the selected sgRNA (20 nt), with the first six nucleotides of sgRNA inversely complementary to the first six nucleotides of the HH-Ribo region (two bold parts of the sequence). The F and R sequences indicate the primers for the fragment synthesis. Sequences with red and green backgrounds indicate recognition sites of the two restriction endonucleases (Nhe I, Bsa I); the broken lines show the cleavage sites, respectively. After double enzymatic digestion of the fragment and vector (the translucent bases indicate the excised portion), the fragment is ligated into the vector to complete the construction of the gene-knockout CRISPR/Cas9 vector

be flanked by HH-Ribo and HDV-Ribo sequences (Fig. 1). When the recombinant plasmid is in the cell, these ribozyme sequences need to be folded and self-cleaved to release the intermediate sgRNA and its scaffold for sgRNA function. The HDV-Ribo sequence is already in the pYF515 plasmid. However, sgRNA primers should be designed, which include the Hammerhead (HH) ribozyme sequence before the selected sgRNA (20 bp), as shown in Fig. 2. Furthermore, in order for the HH ribozyme to form a hairpin structure and for self-cleavage to occur before the release of sgDNA and its scaffold, the first 6 bp of sgRNA needs to be inversely complementary to the first 6 bp of the HH-Ribo region (the two bold parts of the sequence shown in Fig. 2 (*see* **Note 2**) and the whole of the target fragment needs to be flanked by NheI and BsaI sites as in Fig. 2. The whole of the double-stranded target (as shown in Fig. 2) can be synthesized by annealing complementary oligonucleotides. For details, please refer to Fang et al., 2017 [4].

2. Enzyme digestion of pYF515 vector.

- 20 μL 10× Buffer suitable for BsaI and NheI (e.g., Cutsmart)
- 4 μL BsaI
- 4 μL NheI
- 8 μL of the template.

Add distilled water to make 200 μL. Use a DNA Fragment Purification Kit to purify both the digested fragment and the vector.

3. Ligation of sgRNA fragment and the pYF515 vector: 1 μL 10× T4 DNA Ligase Buffer, 1 μL T4 DNA Ligase, 100 ng sgRNA fragment, 100 ng linearized vector. Incubate at 10–16 °C for 2 h.
4. Take 10 μL of the ligation product to transform into *E. coli* JM109 competent cells.
5. Perform colony PCR for rough validation of vector construction, then use a Plasmid extraction Mini Kit to extract the plasmids and send for sequencing with M13F primers.

3.3 Construction of Gene Replacement Vector

In *Phytophthora*, gene replacement has been found to be the most efficient when homologous regions of 1 kb in length are used [1]. In-fusion cloning is an efficient way to achieve the joining of multiple fragments. Here, we use the In-fusion strategy to construct the multi-fragment gene-replacement vector, with 1 kb upstream fragment of the target gene, the replacement gene *HPH* (hygromycin B phosphotransferase gene), and 1 kb downstream fragment sequentially ligated to the cloning vector pBluescript SK II+ (pBS-SK II+) (see Subheading 2, **item 3** part 3). Linearize the vector in a way you are familiar with.

1. For In-fusion cloning, design sets of primers that overlap by 15–20 bases to connect the vector, upstream target gene region, selectable marker/replacement gene (e.g., HPH), and the downstream target gene region.
2. Set up In-fusion cloning (Vazyme) mediated ligation of three fragments with pBS-SK II+ using optimal equimolar amounts as follows:

   Cloning vector: 0.02 ng × number of base pairs.

   Each fragment (left and right flank and HPH): 0.02 ng × number of base pairs.

   5 μL 2× ClonExpress Mix, which contains the recombinase enzyme

   X μL pBluescript SK II+ vector, Linearized.

   Y1 μL upstream1kb fragment.

Y2 μL downstream1kb fragment.

Y3 μL HPH replacement fragment.

~10 μL distilled water.

The volumes (X, Y1, Y2, and Y3) of the vector and fragments are equal to the optimal amount divided by the concentration, and their total volume does not exceed 5 uL.

3. Incubate the reaction in a PCR machine at 50 °C for 15 min, and then hold at 4 °C.
4. Take 10 μL of the ligation product to transform into *E. coli* JM109 competent cells.
5. Perform colony PCR for rough validation of vector construction.
6. Use the FastPure Plasmid Mini Kit to extract the plasmid and send it for sequencing with M13F primers. For the sequence of the HH-Ribo + (sgRNA) + HDV-Ribo (*see* **Note 2**).

3.4 PEG-Mediated Stable Transformation of P. sojae *Protoplasts*

1. Culture the strain to be transformed (e.g., *P. sojae* strain P6497) on NPB-agar plates and incubate in the dark at 25 °C for 3–4 days.
2. Cut 3 × 3 mm mycelial blocks from the edge of the colony with a sharp scalpel and place them in three 250 mL Erlenmeyer flasks containing 50 mL of NPB each. Use 7–8 mycelial blocks per flask and three flasks in total. Incubate the flasks at 25 °C for 2.5–3 days in the dark.
3. Harvest the *P. sojae* hyphae by filtration through autoclaved gauze and place the hyphal mass into a 50 mL centrifuge tube with forceps. Rinse once with about 30–40 mL of 0.8 M mannitol, gently shake to rinse the mycelia, then filter through gauze into a 200 mL beaker to collect the mycelia. Then transfer the rinsed mycelia into a 50 mL centrifuge tube, add 35 mL of 0.8 M mannitol solution, and wash the hyphae on a rotating shaker at 60 rpm for 10 min at room temperature.
4. Collect and add the washed mycelium to a 50 mL centrifuge tube containing freshly prepared enzyme solution. The enzymatic digestion of the mycelia is carried out on a rotating shaker (60 rpm) at room temperature for about 40–50 min (*see* **Note 3**).
5. Use a 50 mL beaker with two layers of sterile Miracloth to filter the mycelium to collect protoplasts. Pour the protoplasts into a 50-mL round-bottom centrifuge tube and centrifuge in a swing-out rotor at 1200 × *g* at 4 °C (in a pre-cooled centrifuge) for 4 min (*see* **Note 4**).

6. Discard the supernatant gently. Add about 10 mL of pre-cooled W5 solution along the wall of the tube to gently resuspend the protoplasts. Then add W5 solution to 35 mL and centrifuge at 4 °C 1200× *g*) for 4 min. Check the protoplast yield using a microscope. The concentration of the protoplast needs to be adjusted to $2x10^6$/ml.
7. Discard the supernatant, add about 10 mL of W5 solution to gently resuspend the protoplasts, leave on ice for 30 min, and then centrifuge at 4 °C and 1200× *g* for 4 min.
8. Discard the supernatant again, then resuspend the protoplasts in pre-cooled MMg solution at a concentration of 2×10^6/mL; leave at room temperature for 10 min.
9. Take an appropriately sized ice box and place it on the ultra-clean bench. Place several new regular 50 mL tubes on ice, label them, and add at least 30 μg of purified plasmid to the bottom of each tube.
10. Add 1 mL of protoplasts to each plasmid-loaded 50 mL centrifuge tube, fully mix by gently tapping the tube, and leave on ice for 10 min.
11. Add 580 μL of 40% PEG solution (see Subheading 2, **item 4** Part 10) to each centrifuge tube by pipetting along the wall of the tube. Repeat two more times, for a total of 1.74 mL PEG solution. Gently turn the centrifuge tube during each addition step, so that the PEG can flow into the mixture of plasmids and protoplasts all around the wall of the tube, and then gently tap the tubes to mix well. Place the tubes on ice for 20 min.
12. Add ampicillin (50 mg/mL) to PM broth on ice to a final concentration of 50 μg/mL and mix well before use.
13. Add 2 mL of PM broth to each tube, slowly invert it once, and place on ice for 2 min.
14. Add 8 mL of PM broth to each tube, slowly invert it once, and place on ice for 2 min.
15. Add 10 mL of PM broth to each tube and incubate in a tilted stand at 25 °C for overnight regeneration.
16. From the overnight growth (~14 h) cultures, examine 5 μL microscopically for regeneration and centrifuge the tubes at 2000× *g* for 5 min to precipitate the regenerated hyphae.
17. Discard the supernatant but leave about 5 mL of liquid remaining in the tube. Gently pipette up and down to suspend the hyphae, then add 35 mL of PM agar containing 25 ug/mL G418. Mix and pour into Petri dishes, blow dry off excess moisture, and incubate at 25 °C in the dark for 3–4 days. The agar needs to be cooled down post-autoclaving but kept liquid, with the temperature of the hot liquid agar maintained at approximately 40–50 °C.

18. After 3–4 days, many colonies should be seen on the surface of the agar; cover the colonies with 25 mL of V8 agar (*see* **Note 5**) containing 25 μg/mL G418, air dry, and continue the incubation at 25 °C in the dark for an additional 3–4 days.
19. When mycelial tufts are growing again through the V8 agar layer, cut a single small, regenerated colony with dense hyphae and transfer it to a V8 agar plate containing 50 μg/mL G418 for further cultivation and screening. It is recommended that resistant *P. sojae* transformants are grown on G418-containing plates for 7 days before screening.

3.5 Screening and Identification of Knockout Mutants

Extract the genomic DNA of the candidate transformants, then screen and verify the knockout mutants by PCR and sequencing. Validated mutants are then grown in antibiotic-free V8 medium for two consecutive generations before phenotypic analysis.

3.5.1 Extract Genomic DNA from Mycelia of Putative Transformants Using a Quick/Crude CTAB Method

1. Scrape an appropriate amount of mycelia (~1 cm × 2 cm area, or larger) into a 2 mL centrifuge tube, then add 600 μL of 2% CTAB and 600 μL of chloroform and place the tube in an incubator shaker at 37 °C for 2 h.
2. Centrifuge the tube at 13660× *g* for 20 min. Pipette the supernatant (~400 μL) into a new 1.5 mL tube. Do not pipette up any of the solid material. Add 800 μL of absolute ethanol (twice the volume of supernatant), mix well, and place it at −20 °C for at least 30 min to precipitate the DNA.
3. Centrifuge the tube at 13660× *g* for 10 min, discard the supernatant, centrifuge again for 2 min, and remove the remaining supernatant with a pipette.
4. Place the centrifuge tube with its lid open in an oven at 50 °C for 30 min. Alternatively, air dry for 1 h or more for complete volatilization of ethanol.
5. Resuspend the DNA pellet in 30 μL sterile distilled water at room temperature and store at −20 °C before use.

3.5.2 Refined Extraction of Genomic DNA with a DNA Extraction Kit

Use a gDNA extraction kit, such as the Tiangen plant genome extraction kit (DP320–50), according to the manufacturer's instructions.

3.5.3 Screening for Knockout Mutants

1. The DNA product from the crude CTAB extraction method is used as the template, and the full-length primers of the target gene (F_1, R_1 in Fig. 3) are used for PCR amplification. Transformants that do not amplify the fragment of the target gene are selected as candidate knockout mutants.
2. The DNA product from the refined kit-based extraction method is used as a PCR template. Perform the PCR using a primer that binds outside the 1 kb left arm (F_2 in Fig. 3) and

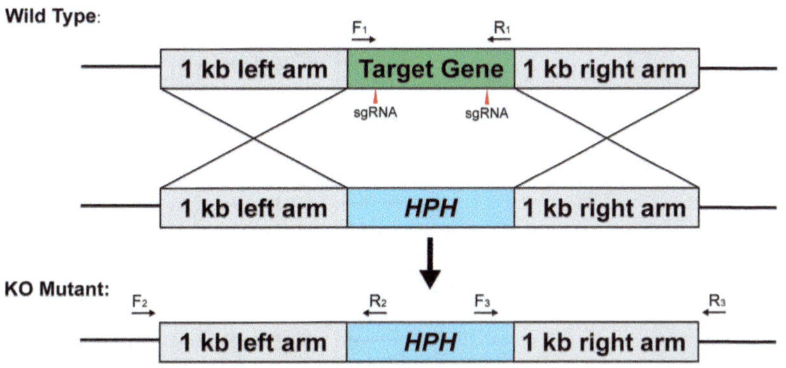

Fig. 3 Homology-directed repair (HDR) mediated gene replacement with primers for mutation screening. Two sgRNAs (two red triangles) targeted at the gene are designed. Primer pairs, F1/R1, F2/R2, F3/R3 can be used for screening HDR mutants

internal primer of the HPH replacement gene (R_2 in Fig. 3) to check the upstream interface of the replacement fragment. Another internal primer of the HPH replacement gene (F_3 in Fig. 3) and a primer outside the 1 kb right arm (R_3 in Fig. 3) are used to check the downstream interface of the replacement fragment to make sure that the replacement fragment is successfully integrated into the correct position in the genome (Fig. 3).

Perform the PCR in a 25 μL volume, as per the steps below. Remember to include a positive control (e.g., wild type of *P. sojae* 6497) and a negative control (distilled water).

(a) Make up a master mix containing for each sample:
- 18.25 μL distilled water
- 2.5 μL 10× PCR Buffer (Mg^{2+} plus)
- 2 μL 1.25 mM dNTP solution
- 0.25 μL forward primer solution (10 μM)
- 0.25 μL reverse primer solution (10 μM)
- 0.25 μL Taq DNA polymerase (5 U/μL).

(b) Distribute the master mix to each PCR tube (23.5 μL per tube) and add 1.5 μL of template DNA per tube.

(c) Amplify DNA using the following PCR programme: Initial denaturation at 95 °C for 5 min followed by 32 cycles of 95 °C for 30 s, × °C for 30 s, 72 °C for y min, and then 72 °C for 5 min and an infinite hold step of 4 °C.

(d) Prepare a 1% horizontal agarose gel. Run 5 μL of the amplification product +1 of 6× μL of loading dye at about 100 V for about 30 min.

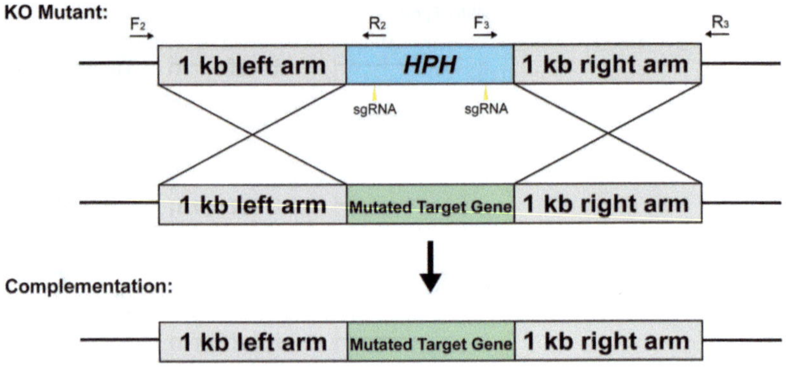

Fig. 4 A schematic representation of the complementation strategy. New sgRNAs (two yellow triangles) targeted at the replacement gene (the HPH gene) are designed. The mutated target gene refers to the original gene of interest but with mutated sgRNA sites

3. To verify the PCR results, check the candidate knockout mutants by sequencing with the general primer M13F.

3.6 Screening and Analysis of sgRNA Targets for in Situ Complementation

Use the same tools and steps as described in Subheading 3, **step 1**. The difference between these two parts is the input sequences: the ORF of the targeted gene sequence is used in Subheading 3, **step 1**, while in this step, the sequence of HPH is used, in order to inactivate the HPH gene (Fig. 4).

3.7 Construction of Replacement HPH Gene Knockout CRISPR/Cas9 Vector for In Situ Complementation

This part uses two sgRNAs targeting the HPH gene to knock out the replacement gene. The methods for sgRNA synthesis and vector construction are the same as in Subheading 3, **step 2**.

3.8 Construction of Complementation Gene Vector

1. Amplify 1 kb of each upstream and downstream sequence of the target gene (same as Subheading 3, **step 3**) and the reintroduction fragment (the original gene that was knocked out but with mutated sgRNA sites) and purify them.
2. The methods for constructing a vector with multiple fragments are the same as in Subheading 3, **step 3**. After culturing the mutants for at least three generations in antibiotic-free plates for more than two weeks, the mutants could no longer grow on G418 antibiotic plates [2]. Thus, the gene knockout strains can be complemented by G418 Resistance. This method is not suitable for the knockout strains that still have G418 resistance.

3.9 PEG-Mediated Stable Transformation of P. sojae Protoplasts for In Situ Complementation

The transformation steps are the same as described in Subheading 3, **step 4**. The difference is that new complementation plasmids are used.

3.10 Screening and Identification of Complemented Mutants

After growing on G418-containing plates for 7 days, extract the gDNA of the candidate complementation transformants using the CTAB method and screen and verify the complementation of the gene knockout by genomic PCR and sequencing methods.

Two rounds of screening are performed on the complemented strains. The first round of PCR amplifies the sequence at the junctions of the flanking sequences and HPH to detect HPH deletion events. A second round of PCRs will detect in situ complementation events using primers located outside the homologous arm and within the ORF of the original knockout gene that has mutated. Validated mutants are activated in antibiotic-free V8 medium for two consecutive generations before phenotypic analysis.

4 Notes

1. The pYF515 empty vector does not contain the sequence of HH-Ribo region, compromising the possibility of using the empty vector as a template, and it is advisable to use any pYF515 that has inserted sgRNA fragment as a template. If there are no available templates, please directly synthesize these fragments.
2. The designed sequence of HH-Ribo + sgRNA + HDV-Ribo should be as follows: 5′-**NNNNNN** CTGATGAGTCCGT GAGGACGAAACGAGTAAGCTCGTC **NNNNNN(N)14** G TTTTAGAGCTAGAAATAGCAAGTTAAAATAAGGC TAGTCCGTTATCAACTTGAAAAAGTGGCACC GAGTCGGTGCTTTTGGCCGGCATGGTCC CAGCCTCCTCGCTGGCGCCGGCTGGGCAA CATGCTTCGGCATGGCGAATGGGA -3′. The Ns here represent any base and "(N)14" means there are 14 bases. The underlined part "**NNNNNN-(N)14**" is the selected sgRNA target (20 bp).
3. Check at 40 min. Enzymatic hydrolysis can be considered almost complete when 80% of the mycelia has disappeared. Extend the enzymatic time appropriately according to the enzymatic digestion.
4. The subsequent operations need to be carried out at low temperatures and all the materials and solutions used need to be pre-cooled to 4 °C.
5. Both PM and V8 agar are suitable, but V8 is cheaper, so V8 is recommended here.

References

1. Fang Y, Tyler BM (2016) Efficient disruption and replacement of an effector gene in the oomycete *Phytophthora sojae* using CRISPR/Cas9. Mol Plant Pathol 17(1):127–139. Epub 2015/10/29. doi:https://doi.org/10.1111/mpp.12318. PubMed PMID: 26507366; PubMed Central PMCID: PMC6638440
2. Qiu M, Li YI, Ye WW, Zheng XB, Wang YC (2021) A CRISPR/Cas9-mediated *in situ* complementation method for *Phytophthora sojae* mutants. Mol Plant Pathol 22(3):373–381. https://doi.org/10.1111/mpp.13028. PubMed PMID: WOS:000610447400001
3. Tyler BM, Tripathy S, Zhang X, Dehal P, Jiang RH, Aerts A et al (2006) *Phytophthora* genome sequences uncover evolutionary origins and mechanisms of pathogenesis. Science 313(5791):1261–1266. Epub 2006/09/02. https://doi.org/10.1126/science.1128796
4. Fang Y, Cui L, Gu B, Arredondo F, Tyler BM (2017) Efficient genome editing in the oomycete *Phytophthora sojae* using CRISPR/Cas9. Curr Protoc Microbiol 44:21A 1 1–21A 1 6. Epub 2017/02/07. https://doi.org/10.1002/cpmc.25

Chapter 6

Fluorescence In Situ Hybridization Protocol for Visualization of Oomycetes In Vitro and In Planta

Glen Groben, Catalina Salgado-Salazar, and Jo Anne Crouch

Abstract

Fluorescence in situ hybridization enables the visualization of organisms in the environment without having to culture them. Here, we describe a FISH protocol to visualize oomycete structures (mycelia, sporangiophores, sporangia, and oospores) directly as well as from colonized plant material. The protocol utilizes organic compounds with low toxicities and does not require a permeabilization step. The protocol was used to visualize impatiens downy mildew *Plasmopara destructor* sporangiophores colonizing an *Impatiens walleriana* leaf.

Key words FISH, Fluorescence in situ hybridization, Culture-independent visualization

1 Introduction

Fluorescence in situ hybridization (FISH) is a commonly used method to visualize nucleic acids in the environment, allowing researchers to study organisms without having to culture them. The technique is used in clinical diagnostics, epidemiology, bacteriology, virology, microbial ecology, plant pathology, and many other fields. FISH was first described in 1969 [1, 2] using radiolabeled probes, but over time the method has progressed to the use of fluorescent-labeled probes due to their ease of use [3]. FISH is based on four core steps: fixation of the sample, permeabilization to increase the accessibility of the probe, hybridization of the probe, and washing unbound probe [4]. The probe is usually a 15 to 30 base pair oligonucleotide that is specific to the organism(s) of interest and contains a base pair tagged with a fluorophore [5]. Ribosomal rRNA genes are often used as targets for FISH probes because of their high copy numbers in the cell and the fact that these targets typically contain variable nucleotide sequences that provide high levels of specificity [6]. Here we describe a FISH procedure [7] that utilizes organic compounds with low toxicity

Monica L. Gerth and Rosie E. Bradshaw (eds.), *Phytophthora: Methods and Protocols*, Methods in Molecular Biology, vol. 2892, https://doi.org/10.1007/978-1-0716-4330-3_6,

levels [8–10] and does not require an additional permeabilization step [4]. The protocol outlined here targets *Plasmopara destructor* (previously known as *Plasmopara obducens*), an oomycete that causes downy mildew disease of *Impatiens walleriana* [7] but could be modified to target any oomycete by changing the oligonucleotide probe sequence.

2 Materials

2.1 General Equipment

1. Personal Protective Equipment: safety goggles, lab coat, gloves.
2. Microscope or stereomicroscope with fluorescence capabilities, camera and software. This protocol was developed using a Zeiss Axio Zoom V16 equipped with a HXP200C fluorescence light source, a 63 HE filter (excitation BP 572/25, emission 629/62), an Axiocam 705 Mono digital camera and Zen 3.6 blue edition software (Carl Zeiss).
3. Drying oven that can incubate samples at two different temperatures or two drying ovens, such as Hybaid H9270 Dual Stack Rotisserie Laboratory Hybridization Incubator Oven.
4. Glass microscope slides and cover slips (*see* **Note 1**).
5. Antifade Mounting Media like ProLong® Diamond Antifade Mountant (ThermoFisher).

2.2 Sample-Size Dependent Equipment and Materials for FISH Assays

Depending on the size of the sample that is being subjected to the FISH assay, one of the following sets of equipment and materials will be required (*see* **Note 2**).

Option 1: Visualization of smaller objects (mycelia, sporangiophores, sporangia, oospores, etc.):

1. Hybridization chambers with 65 μL capacity (e.g., Frame-Seal™ incubation chambers, BIO-RAD Laboratories).
2. Hotplate.
3. 50 mL conical polypropylene tubes.

Option 2: Plant tissue (leaf, stem, root, etc.):

1. 35 mm × 10 mm sterile polystyrene Petri dishes (*see* **Note 3**).
2. Airtight container, such as Ziplock© square containers (709 mL = 1.5 pints).
3. Refrigerated area set to 4 °C.

2.3 Preparing Stock Solutions

Depending on the sample type, different volumes of buffer will be needed. For plant tissue such as leaves, stems, or roots, 10 mL of the ethanol and wash buffer solutions is enough for one sample. The volume of the solutions will vary for the smaller objects depending on the location of the Frame-seal™ (*see* **Note 4**) and how many hybridization chambers are on one microscope slide. Roughly 20 mL will cover 1/3 of the slide, 30 mL will cover 2/3 of the slide and 40 mL will cover the entire slide in a 50 mL conical tube.

1. Ethanol Dilutions 70%, 85%, and 96%: Using 100% ethanol, add 7, 8.5, and 9.6 mL to separate 15 mL tubes. Using PCR-grade water add 3 mL, 1.5 mL, and 400 μL respectively for a total of 10 mL of each dilution. Ethanol is flammable and should be stored in a flammable storage cabinet at room temperature.
2. Hybridization wash buffer 1: 0.05 M Tris–HCl pH 7.6, 0.3 M NaCl, 0.1% Tween 20. To prepare 10 mL, using a 15 mL tube add 500 μL of 1 M Tris-HCl, pH 7.6, 0.175 g NaCl, 10 μL Tween 20, and bring to a final volume of 10 mL using PCR-grade water.
3. Hybridization wash buffer 2: 0.05 M Tris–HCl pH 7.6, 0.15 M NaCl, 0.1% Tween 20. To prepare 10 mL, using a 15 mL tube add 500 μL 1 M Tris-HCl 1 M pH 7.6, 0.0877 g NaCl, 10 μL Tween 20, and bring to a final volume of 10 mL using PCR-grade water.

   A smaller volume of 1 mL is needed for the hybridization buffer that should be made in an opaque tube, or a tube covered with aluminum foil (*see* **Note 5**).
4. Hybridization Buffer: 20% 50% w/v dextran sulfate, 15% ethylene carbonate, 600 mM NaCl, 10 mM citrate buffer pH 6.2, 2 ng L^{-1} fluorescent probe. To prepare 1 mL, using a 1.5–2 mL tube add 200 μL dextran sulfate 50% solution (w/v) (*see* **Note 6**): 150 μL ethylene carbonate (*see* **Note 7**): 111 μL-citrate buffer 0.09 M pH 6.2, and 0.035 g NaCl. Bring the final volume to 1 mL with PCR-grade water. Store at room temperature. The 2 ng of probe should be added right before beginning the hybridization step to minimize fading of the fluorescence probe (*see* **Note 8**).

   The following materials (i.e., fixation buffer, fixation clearing buffer, and 1× phosphate buffered saline (PBS)) are only needed if using plant tissue.
5. Fixation Buffer 4% formaldehyde: Prepare and store in a flammables storage cabinet at room temperature. To a 15 mL tube add 1.08 mL 37% formaldehyde and 8.92 mL PCR grade water for a total of 10 mL. Formaldehyde is flammable and should be stored in a flammables storage cabinet at room temperature.

6. Fixation Clearing Buffer: 75% Ethanol 95%, 15% acetic acid, 10% glycerol. Prepare and store in a flammables cabinet at room temperature. In a 15 mL tube add 7.5 mL ethanol 95%, 1.5 mL acetic acid, and 1 mL glycerol for a total of 10 mL. Ethanol and acetic acid are flammable and should be stored in a flammables cabinet at room temperature.
7. 1× Phosphate-buffered saline (PBS) pH 7.4: 8 g NaCl, 0.2 g KCl, 1.44 g Na_2HPO_4, 0.245 g KH_2PO_4, PCR-grade water to 900 mL. Adjust pH to 7.4 and water to a final volume of 1 L. Alternatively, PBS solutions can be purchased from chemical companies like ThermoFisher Scientific, Sigma-Aldrich and many more. The purchased PBS may come in higher concentrations of 10× or 20×, which will need to be diluted to 1×.

3 Methods

3.1 Probe Design

The sequence of the oligonucleotide probe will determine which organism(s) will be visualized by the FISH assay. The probe is synthesized with a fluorophore-tagged base pair anywhere on the probe but usually on the first or last base pair [6]. Make sure that the fluorophore emission can be detected by the microscope with fluorescence capabilities. Background autofluorescence is common, and wavelengths in these regions should be avoided. Chlorophyll autofluorescence is in the green spectra of light, making it more difficult to detect fluorophores that emit green light, so the use of green fluorophores to label the probe is not recommended if plant tissue will be part of your sample. More than one probe can be used to image multiple organisms at the same time [11] and non-fluorescent tagged probes can be used to reduce off-target hybridization [12].

Designing a probe follows a few simple steps: obtaining the relevant DNA sequence data, aligning the sequences, selecting the probe region, and checking for secondary structures.

1. Sequencing information can be obtained either by sequencing DNA or downloading sequences from a publicly available database like NCBI GenBank. Collect sequences for the organism of interest as well as for a cohort of closely related taxa to avoid off-target amplification when designing the probe.
2. Align the sequences using a multiple sequence alignment tool like MUSCLE or MAFFT or using commercial software such as Geneious or CLC Genomics Workbench. Make sure that the sequences are all in the same direction (5′ to 3′ or 3′ to 5′) before running the multiple sequence alignment tool.
3. Use the sequence alignment to help select a specific probe region. The ideal length of the probe is around 15–30 bp long [5]. Select regions in the multiple sequence alignment

that are specific to the target organism(s) and dissimilar to the non-target organisms. Look for regions that have a high number of gaps and mismatches between the target and non-target organisms.

4. Use a bioinformatics tool such as the IDT OligoAnalyzer or Sigma OligoEvaluator, to check that the probe sequences will not bind to themselves (with a secondary structure calculator) and that the GC content is between 35% and 65%.

3.2 Sample Collection

Collect a sample of interest and place it into the appropriate carrier (i.e., microscope slide or Petri dish). For smaller samples like mycelia, oospores, sporangia, or zoospores, place the sample on a glass microscope slide, taking care not to overload the slide with too much material. For larger samples like leaves, stems, or roots, use a 35 mm × 10 mm sterile polystyrene Petri dish.

3.3 Fixation of Small Samples

Depending on the sample of interest, two different fixation protocols can be used.

1. For small samples (mycelia, sporangiophores, sporangia, oospores, etc.), place the glass microscope slide containing the sample on a 60 °C hotplate for 15 s.
2. After incubation, place a 65 μL capacity hybridization chamber (e.g., Frame-Seal™) around the sample.
3. Proceed to the hybridization step (Subheading 3.5).

3.4 Fixation of Larger Samples

For larger, macroscopic samples such as plant leaves, stems, or roots infected with the oomycete target, conduct the following steps in a chemical cabinet.

1. Open the Petri dish containing the plant sample and pipette enough 4% formaldehyde solution to submerge the sample (*see* **Note 9**). Close the Petri dish and place it inside an airtight container. Incubate the samples overnight at 4 °C.
2. After incubation, bring the samples back to the chemical hood. Remove the 4% formaldehyde solution by pipetting and place it into an appropriate waste container. Wash the samples with 1× PBS solution (*see* **Note 10**).
3. Remove the 1× PBS solution by pipetting and wash again with a second round of 1× PBS solution.
4. Remove the 1× PBS solution by pipetting and wash with the fixation-clearing buffer.
5. Remove the fixation buffer by pipetting and place it into an appropriate waste container.
6. The samples can be stored in 50% v/v 1× PBS, 50% v/v ethanol 100% solution at 4 °C for up to 6 months.

3.5 Hybridization

1. Preheat the drying oven(s) to 67 °C and 45 °C. Aliquot 1 mL of hybridization buffer into a 1.5–2 mL tube. Heat the tube on a heat block or water bath to 40 °C (*see* **Note 5**).
2. The amount of solution used for the following dehydration and wash steps will depend on the sample size. Unless stated otherwise, the dehydration and wash steps should submerge the sample. General instructions for all subsequent wash steps are given in **step 3** (for smaller samples) and **step 4** (for larger samples).
3. For smaller samples, the entire microscope slide should be moved via forceps into the different wash buffer tubes set up in 50 mL conical tubes. Ensure that enough buffer is added to submerge the microscope slide.
4. For larger (e.g., plant) samples, the buffers should be added and removed from the Petri dish by pipetting around 3–4 mL of the different buffers.
5. Dehydrate the samples with a series of increasingly concentrated ethanol washes at room temperature (20–25 °C) as follows. Wash the samples with 70% ethanol, using the methods described in **step 3** (for smaller samples) or **step 4** (for larger samples), and wait 2 min. Then, wash with 85% ethanol and wait 2 min. Finally, wash with 96% ethanol for 2 min, remove the buffer from the sample, then air dry the sample for approximately 5 min.
6. Remove the hybridization buffer from the heat block, then add 2 ng of probe to the hybridization buffer (*see* **Note 8**). Vortex the hybridization buffer/probe solution.
7. Add hybridization buffer/probe solution to the samples. To avoid excess background fluorescence, add only a small amount: ~50 μL for microscope slides and ~ 200 μL for samples in Petri dishes. For a microscope slide, pipette the hybridization probe solution into one end of the Frame-seal™, place the flexible plastic Frame-seal™ coverslip at the same end, roll it over the sample, and use a blunt object to rub out the air bubbles. For Petri dishes, pipette the ~200 μL directly on the sample, place a glass coverslip at one end, slowly lower the cover slip, and use a blunt object to rub out the air bubbles. The probe will be excited by light sources, so work quickly to minimize the amount of time the probe is exposed to light (*see* **Note 11**).
8. Incubate the sample in a dark drying oven at 67 °C for 10 min.
9. Place the sample in a dark 45 °C drying oven and incubate for 1 h.
10. After the 1 h incubation, change the temperature in the 67 °C drying oven to 65 °C for the following wash step. Remove the

coverslip and wash the sample with hybridization wash buffer 1 as described in **step 3** (for smaller samples) or **step 4** (for larger samples), then incubate in the drying oven at 65 °C for 10 min.

11. Remove the sample from the oven. Wash the sample with hybridization wash buffer 2 as described in **step 10**. Incubate at room temperature (20–25 °C for 3 min). Wash the sample a second time with hybridization wash buffer 2. Incubate at room temperature (20–25 °C) for 3 min.
12. Dehydrate the samples with a series of increasingly concentrated ethanol washes at room temperature (20–25 °C for 3 min). Wash the samples as described in **step 5** with 70% ethanol and wait 2 min. Wash with 85% ethanol and wait 2 min. Wash with 96% ethanol for 2 min, remove the buffer from the sample, and air dry the sample.
13. For small samples on microscope slides, remove the frame-seal from the microscope slides, add a drop of ProLong® Diamond Antifade Mountant (ThermoFisher, Carlsband, CA, USA), and place a glass coverslip on the slide. For samples in a Petri dish, transfer the sample to a glass microscope slide, add a drop of ProLong® Diamond Antifade Mountant directly onto the sample, and place a glass coverslip on the slide. Cure the samples overnight at 4 °C in the dark.
14. The samples are now ready to be imaged with a microscope or stereomicroscope equipped with fluorescent imaging capabilities.

3.6 Imaging

The steps involved in imaging the sample will vary depending on the equipment used. The following steps were performed on a Zeiss Axio Zoom V16 equipped as described in Subheading 2.1, **item 2**. The autoexposure time was calculated by the software and used to avoid oversaturation. The software was used to collect 15–20 Z-stack images, and a 2D final image was generated using the extended depth of focus setting. The hybridization probe signal was assigned the color violet, and a scale bar was added (Fig. 1). Users will need to optimize the imaging steps based on their own equipment specifications and capabilities.

4 Notes

1. Adhesive glass slides can also be used to help keep the structures in place and are commonly used with soil samples.
2. The fixation step will depend on the size of the samples. Smaller samples that fit on a microscope slide easily can be heat-fixed

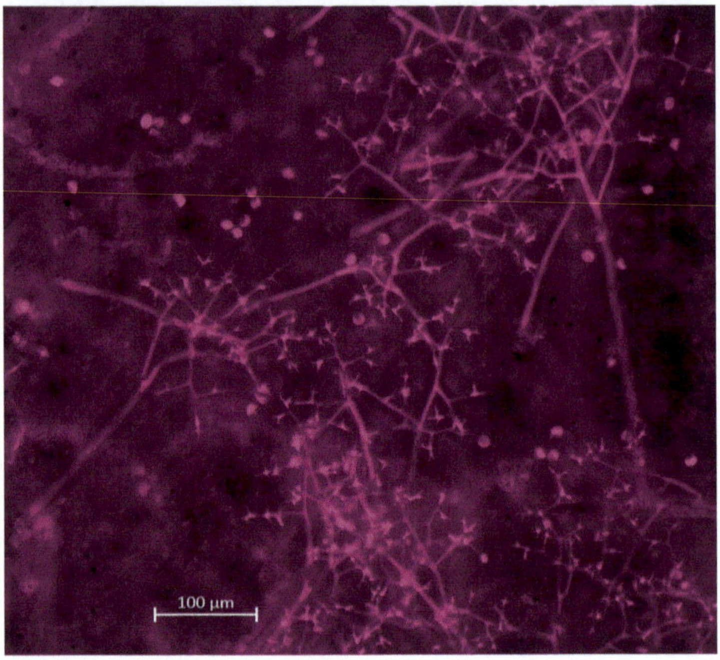

Fig. 1 *Plasmopara destructor* sporangiophores colonizing *Impatiens walleriana* leaf visualized with Alexa594 fluorophore probe labelled as the color violet

while larger samples will need to be chemically fixed in a formaldehyde solution overnight.

3. Larger sized Petri dishes can be used, but more reagents will be required.
4. If the frame seal is placed in the middle of the slide, then 30 mL will be needed to cover it. If the frame seal is placed on the edge of the slide, then only 20 mL is needed to cover the frame seal.
5. The hybridization probe is light-sensitive and needs to be protected from light.
6. Dextran sulfate can be prepared using 5 g dextran sulfate to 10 mL water or purchased as 50% (w/v) solutions.
7. Ethylene carbonate is a solid at room temperature and a liquid at temperatures above 37 °C. In a chemical hood, transfer some solid ethylene carbonate into a heat-stable tube.
8. The probe comes in a dry powder and needs to be reconstituted in nuclease-free water. The volume of water will depend on the concentration of synthesized probe. A recommended water volume is typically suggested by the probe supplier.
9. Approximately 3–4 mL should be added to fully submerge the plant tissue in the 35 mm × 10 mm Petri dishes. Bigger samples and larger Petri dishes will require more solution.

10. All wash steps with plant tissue used approximately 3–4 mL by adding it directly to the Petri plate, waiting a couple of seconds, then removing the fluid.
11. Once the probe is added, everything must be done in dark conditions. Cover the hybridization buffer tube with aluminum foil, cover the windows of the drying oven, do the hybridization wash buffer incubations and dehydration steps inside a box or inside a drawer, and cover the microscope.

Acknowledgements

The authors acknowledge Dr. Margery Daughtrey, Senior Extension Associate at Cornell University, for sharing infected *Impatiens walleriana* leaves used to produce Fig. 1.

References

1. Gall JG, Pardue ML (1969) Formation and detection of RNA-DNA hybrid molecules in cytological preparations. Proc Natl Acad Sci USA 63:378–383
2. John HA, Birnstiel ML, Jones KW (1969) RNA-DNA hybrids at the cytological level. Nature 223:582–587
3. Bauman JG, Wiegant J, Borst P, van Duijn P (1980) A new method for fluorescence microscopical localization of specific DNA sequences by in situ hybridization of fluorochrome labelled RNA. Exp Cell Res 128:485–490
4. Frickmann H, Zautner AE, Moter A, Kikhney J, Hagen RM, Stender H, Poppert S (2017) Fluorescence in situ hybridization (FISH) in the microbiological diagnostic routine laboratory: a review. Crit Rev Microbiol 43:263–293
5. Amann R, Fuchs BM (2008) Single-cell identification in microbial communities by improved fluorescence in situ hybridization techniques. Nature Rev Microbiol 6:339–348
6. Moter A, Göbel UB (2000) Fluorescence in situ hybridization (FISH) for direct visualization of microorganisms. J Microbiol Methods 41:85–112
7. Salgado-Salazar C, Bauchan GR, Wallace EC, Crouch JA (2018) Visualization of the impatiens downy mildew pathogen using fluorescence in situ hybridization (FISH). Plant Methods 14:1–9
8. Matthiesen SH, Hansen CM (2012) Fast and non-toxic in situ hybridization without blocking of repetitive sequences. PLoS One 7: e40675
9. Moffitt JR, Hao J, Wang G, Chen KH, Babcock HP, Zhuang X (2016) High-throughput single-cell gene-expression profiling with multiplexed error-robust fluorescence in situ hybridization. Proc Natl Acad Sci USA 113:11046–11051
10. Wahl GM, Stern M, Stark GR (1979) Efficient transfer of large DNA fragments from agarose gels to diazobenzyloxymethyl-paper and rapid hybridization by using dextran sulfate. Proc Natl Acad Sci USA 76(8):3683–3687
11. Kong Y, Nielsen JL, Nielsen PH (2005) Identity and ecophysiology of uncultured actinobacterial polyphosphate-accumulating organisms in full-scale enhanced biological phosphorus removal plants. Appl Environ Microbiol 71:4076–4085
12. Stender H, Mollerup TA, Lund K, Petersen KH, Hongmanee P, Godtfredsen SE (1999) Direct detection and identification of *Mycobacterium tuberculosis* in smear-positive sputum samples by fluorescence in situ hybridization (FISH) using peptide nucleic acid (PNA) probes. Int J Tuberc Lung Dis 3:830–837

Chapter 7

Rapid and Robust Monitoring of *Phytophthora* Infectivity Through Detached Leaf Assays with Automated Image Analysis

Hannah F. Robinson and Jochem N. A. Vink

Abstract

The detached leaf assay is a valuable method for studying plant-pathogen interactions, enabling the assessment of pathogenicity, plant resistance, and treatment effects. In this protocol, we outline how to set up a *Phytophthora* detached leaf assay and use non-expert machine learning tools to increase the reliability and throughput of the image analysis. Utilizing ilastik for pixel classification and Python scripts for segmentation, manual correction, and temporal linking, the pipeline provides objective and quantitative data over time. The protocol covers assay setup and image segmentation and outlines key considerations, providing a comprehensive guide for setting up and analyzing detached leaf assays. The very minimal material requirements and user-friendly software make this protocol accessible for all *Phytophthora* researchers.

Key words Detached leaf, *Phytophthora*, Machine learning, Image segmentation

1 Introduction

The detached leaf assay is a semi-high throughput, laboratory-based method for evaluating plant-pathogen interactions. It is performed by inoculating detached leaves with a pathogen and monitoring the spread of infection over time, either visually or by image analysis [1]. This is done to assess pathogenicity, plant resistance, and potential treatments. For instance, detached leaf assays are used to screen wild plant species and cultivars for susceptibility to infection by *Phytophthora* species in order to advance resistance breeding. Detached leaf assays are also used to determine the protective and curative properties of chemical fungicides and biological control agents, such as bacteria and fungi [2–11]. Compared to whole-plant assays or field trials, detached leaf assays allow for the rapid assessment of numerous pathogen isolates and plants, as these assays are inexpensive, not laborious due to a simple setup

Monica L. Gerth and Rosie E. Bradshaw (eds.), *Phytophthora: Methods and Protocols*, Methods in Molecular Biology, vol. 2892,
https://doi.org/10.1007/978-1-0716-4330-3_7,

and reduced timeframe, and only require small amounts of space. These assays can be performed in a laboratory setting with common laboratory supplies, avoiding the need for a greenhouse, field testing sites, or quarantine facilities that are unavailable for many researchers. Studies have shown a good correlation between detached leaf assays and field test results [11, 12], which positively indicates the translatability of these laboratory tests to real-world applications.

An additional advantage of detached leaf assays is that disease progression is quantifiable, since the spread of necrosis on the leaf may be measured. With increasing replicates and samples, these measurements become very time-consuming when performed manually, but automated image analysis may be employed to drastically reduce the amount of time required for results analyses. Several automated image analysis tools already exist to detect necrotic spread in images [13–16]. However, most of these tools are difficult to train and use for those with little computer science background. Furthermore, almost all tools are made for the detection of disease in the field, whereas we are specifically interested in applying this technique to the detached leaf assay, which is more controlled and can be used to extract more quantifiable data in regard to growth rate. We developed a pipeline (Fig. 1) that uses ilastik [17], which is software that classifies pixels in an image using random forest models. The software is very easy to use and only requires sparse user annotation that is added interactively, resulting in a rapidly adaptable training set that can be trained on normal desktop computers. The output of the pixel classifications is then run through a set of Python scripts that turn it into segmented objects, allow for manual correction of segmentation mistakes, and link images and leaves within images over time. Altogether, this pipeline allows for an objective, quantitative, high-throughput, and user-friendly determination of necrotic development throughout leaves over time.

2 Materials

2.1 Detached Leaf Assay

1. Freshly harvested leaves.
2. *Phytophthora* species of interest: mycelial plate or zoospores (>10,000/mL).
3. Sterile Petri dishes.
4. Paper towels: sterilized via autoclaving.
5. Scalpel: flame-sterilized.
6. Treatment at the desired concentration in a spray bottle (if required).

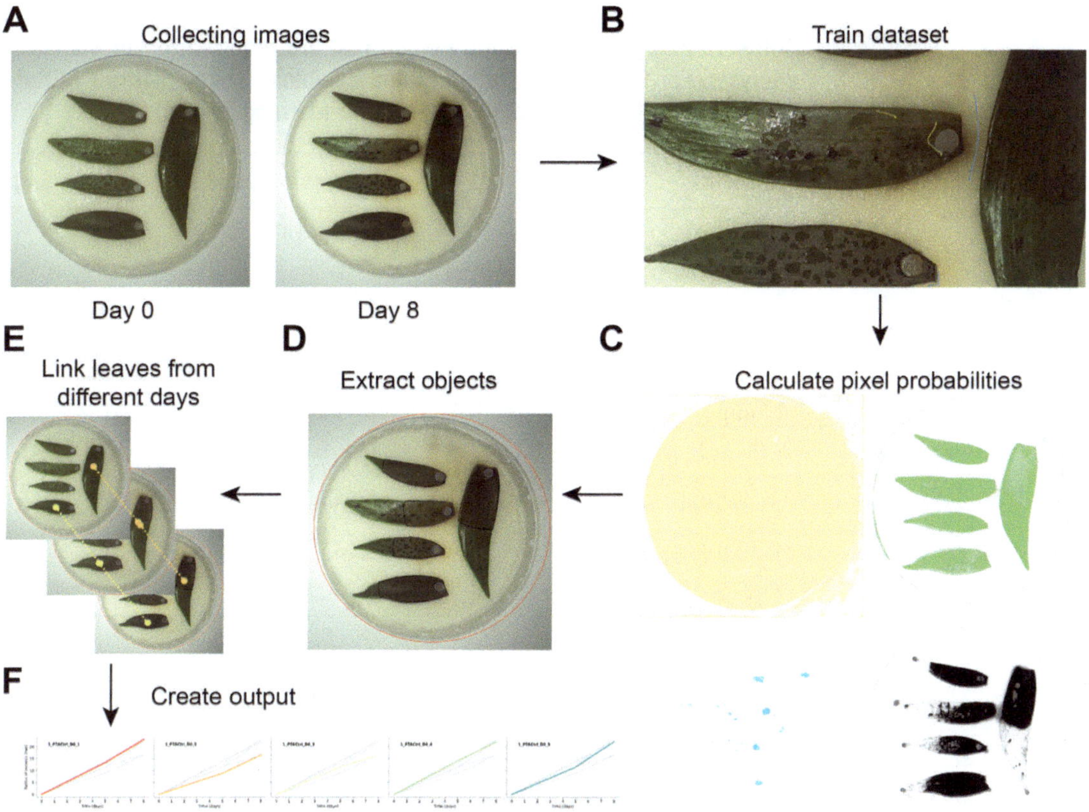

Fig. 1 Overview of the workflow. (**a**) Photos of kauri (*Agathis australis*) leaves exposed to *Phytophthora agathidicida* propagules are taken over multiple days. (**b**) A subset of the images is used to train the dataset in ilastik. (**c**) Subsequently, the image pixels are classified using four different training projects: Petri dish (top left), leaves (top right), plugs/start (bottom left), necrosis (bottom right). (**d**) Based on these pixel classifications, objects are delineated, facilitating the extraction of key information, including leaf and necrosis sizes and the distance from the infection start to the necrosis edge. (**e**) Then, leaves are linked from images taken on different days. (**f**) These linked datasets can then produce output graphs plotting necrotic development over time

7. Cork borer: 3–6 mm diameter, flame-sterilized.
8. Camera.

2.2 Segmentation of Detached Leaf Images

See Table 1.

3 Methods

3.1 Detached Leaf Assay Setup

1. Place the freshly harvested leaves in a Petri dish (*see* **Note 1**) outfitted with two layers of sterilized paper towel moistened with sterile, distilled, deionized water to prevent the leaves from drying out.

Table 1
Software and packages within Python that are required to run the analysis. For the latest installation and download links, visit our GitHub page (https:/github.com/MMElab/Machine-learning-leaves)

Software	Version used at time of writing
Ilastik	v 1.4.0
Spyder (or another python IDE)	Python v 3.9
Python packages	
Pandas	1.3.4
Numpy	1.20.3
Scikit-learn	0.24.2
Matplotlib	3.4.3
H5py	3.2.1
Opencv-python	4.5.5.64

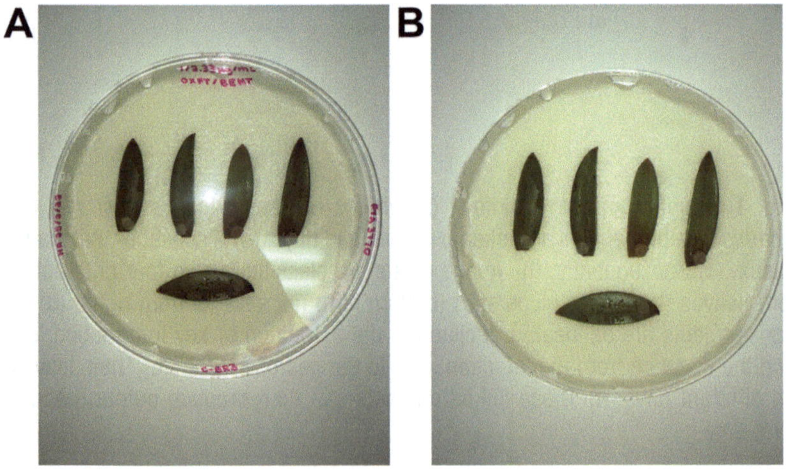

Fig. 2 Detached leaf assay example data (**a**) with lid, (**b**) without lid. The photo with lid is only required for the first timepoint and can be used to retrospectively determine the conditions used for that specific sample; the photo without lid is further processed and analyzed to determine necrotic spread

2. Use a sterilized scalpel to make a 2–3 mm incision 3–5 mm from the base of each leaf (our analysis pipeline only allows for a single point infection per leaf).
3. For assays evaluating the ability of a treatment to protect against infection, spray coat each leaf with the treatment at the desired concentration immediately before inoculation. For assays evaluating the ability of a treatment to cure infection, apply the treatment 3 days post-inoculation (*see* **Note 2**).

4. *Mycelial inoculation:* Use a sterilized cork borer to take a 3–6 mm diameter agar plug (keep plug size consistent) from the leading edge of mycelial growth and position it over the incision, with the mycelium side facing down towards the leaf (*see* **Note 3**). OR—*Zoospore inoculation:* Prepare a suspension of 50,000 zoospores/mL and apply a 20 μL aliquot (i.e., 1000 zoospores total) to the incision (*see* **Note 4**).
5. Take pictures of each Petri dish: first with the labelled lid on and then with the lid off (Fig. 2; *see* **Note 5**). The first picture will be used to identify the sample, and the second picture will be used for data analysis. The image with the lid is only required at the day 0 timepoint, as later timepoints will be linked to day 0 based on leaf morphology.
6. Incubate the plates at 22 °C in a 12 h light:12 h dark cycle until the necrosis reaches ≥70% leaf length. For monitoring of necrotic spread kinetics, pictures should be taken at a regular time interval (e.g., once a day). Alternatively, take pictures at only the start and the end of the experiment if it is necessary to limit the amount of data.

3.2 Training Image Segmentation

1. The images are now ready to be classified with ilastik. For more information on how to use this software, we recommend going through the documentation on https://www.ilastik.org/documentation/.
2. Organize the images into subfolders, where each subfolder is a timepoint of all the samples. Name each subfolder DX, where X is the number of days since inoculation. An example of the data structure we use is provided on the GitHub page (https://github.com/MMElab/Machine-learning-leaves). Only include images of plates without lids in these subfolders.
3. The processing of images is faster when images are in the hdf5 format; therefore, all images are converted into that format. There is a Python script called *h5images.py* that can be run on the datafiles and automatically converts all the images into the correct format. This will create new subfolders called DX_h5, where X is the number of days since inoculation.
4. Open ilastik and create four new pixel classification projects, one for each category: leaf, plug, necrosis, and Petri dish (*see* **Note 6**). Alternatively, previous projects (including the projects in our example folder) can be adapted to new datasets by removing all images from a project. This will ensure the correct settings are already selected in the project file.
5. Load the hf5 images into each project (*see* **Note 7**) at *1. Input Data*. Select all images with Ctrl+A and right-click and select *Edit shared properties*. Set storage to *Copy into project file* for the files to be linked to the ilastik project file (which allows sharing of the training sets with others).

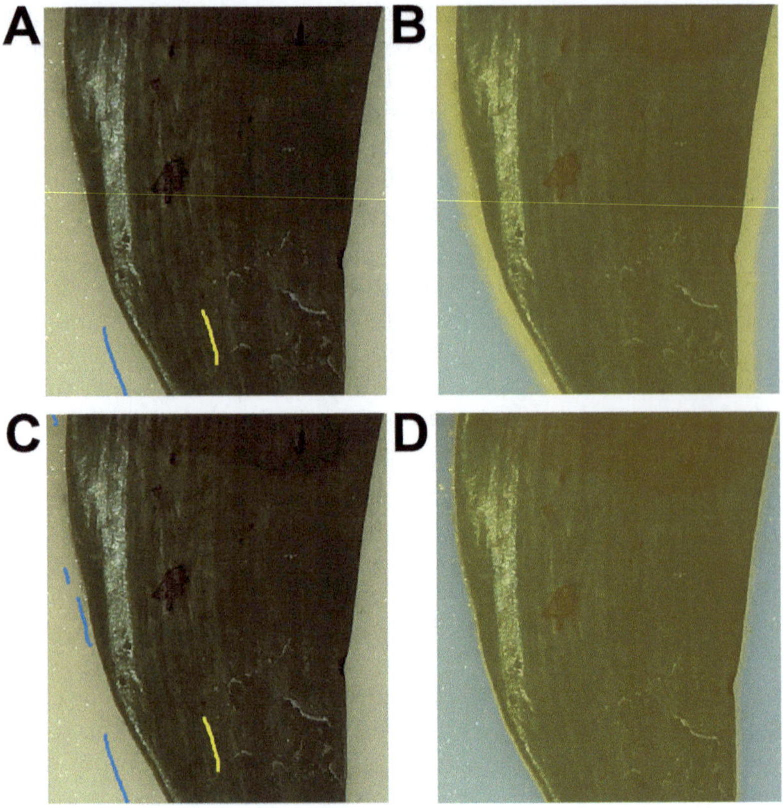

Fig. 3 Training pixel classification. (**a**) The labels drawn into the figure for classification of leaf pixels (yellow: leaf, blue: not leaf). (**b**) The overlaid pixel classification based on the provided labels. (**c**) Labels added to correct overestimation of leaf pixels (**d**). The overlaid pixel classification based on correction labels

6. At *2. Feature Selection*, select all features of several values of σ. This value represents the σ (number of pixels) of the Gaussian that is used to smoothe the image before implementing certain filters. Filters with greater σ values can consequently incorporate information from broader neighborhoods, yet they tend to diminish the finer details through averaging. The number of σ values that should be included will depend on the size of the image/leaves and the amount of computing power available, but in general, we recommend a range that includes 1, 2, 5, 10, and 30 pixels. The more features chosen, the slower the analysis will be. However, the algorithm will have more features from which to train and will therefore be more accurate.
7. At *3. Training*, use Label 1 for pixels that contain the category of interest and Label 2 for pixels that do not contain that category. We recommend using a brush size between 1 and 5 pixels to maintain precision while drawing.

8. Start labelling a small number of pixels in the first image with Label 1 (where that category, such as a leaf, is present) and Label 2 (where that category, such as a leaf, is not present) (*see* **Note 8**). Do not draw too much within a single image: that will slow down the training dramatically and result in overfitting.
9. Move to the next image. Click *Live Update* to see the current prediction ilastik generates.

   Correct the pixels by adding more labels where this prediction is wrong (Fig. 3). Alternate and aim to divide the amount of labelling evenly between the two labels (*see* **Note 8**).
10. Keep going until any unannotated images have a satisfactory classification based on the live prediction generated by ilastik. Potentially revisit earlier images if more training is required.
11. At *4. Prediction Export,* change *Source* to *Probabilities* and click on *Choose Export Image Settings* to change the following default settings:

    Format: compressed hdf5.

    File: "{dataset_dir}/Pixelprobabilities_[**category**]/{nickname}_{result_type}.h5".

    For category, insert necrosis, Petri, leaf, or plug, depending on the project.

    Dataset: exported_data.
12. Save each project file in a separate subfolder called Projects.

3.3 Run Image Segmentation and Further Analysis

1. Open Python software.
2. Run *ilastikbatch.py.* This will prompt the user to select the main folder containing all the images. Then, it will prompt the user to select the folder containing the ilastik project files. If the naming convention of the projects is different from the standard naming, these can be adjusted in the script. Depending on the size of the dataset, this can be a slow and storage-intensive run and might require usage of a high-performance computing cluster. After running this code, extra subfolders that contain the *Pixelprobabilities* files generated by ilastik should have been added to each *DX_h5* folder.
3. Run *extractinfo_ilastikbatch.py.* This will again prompt the user to select the main folder containing all the images. This program will extract features from the pixel classifications and turn them into objects. After running this code, several new files and folders will have been created. Within each *DX_h5* folder, every image file now also contains a number of *_classification.npy* files. These files contain the objects for each category (*see* **Note 9**). Furthermore, we create an *Imageoutput.csv* file, which contains the following columns of quantitative data extracted from each image:

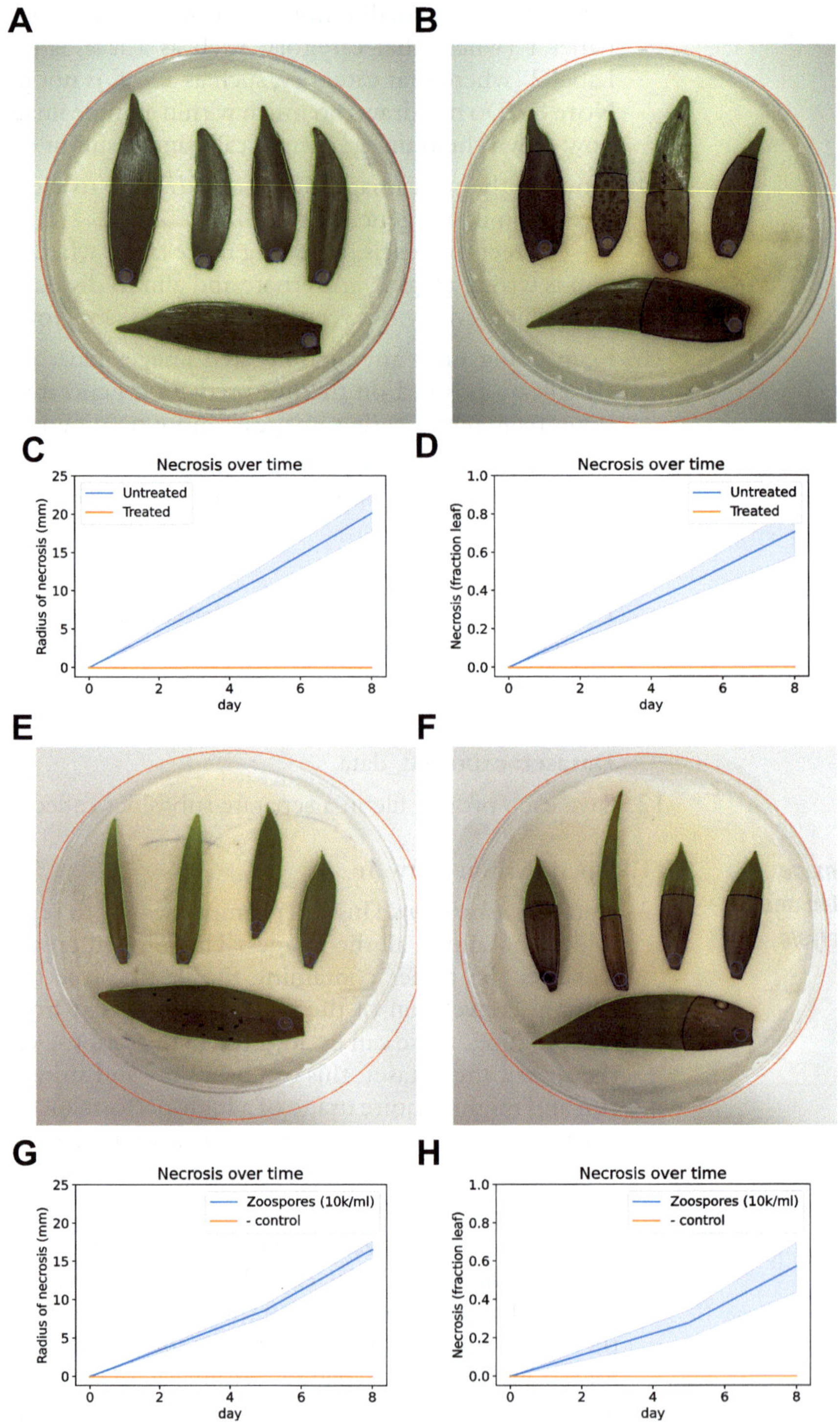

Fig. 4 (**a**) An example of untreated kauri leaves at the start of an experiment (day = 0). All leaves were inoculated with mycelial plugs containing *P. agathidicida*. (**b**) An example of untreated kauri leaves at the end of an experiment (day = 8). All leaves were inoculated with mycelial plugs containing *P. agathidicida*. (**c**) Example plot of the radius of necrosis (mm) over time in infected kauri leaves, comparing untreated (blue) and Zorvec Enicade-treated (orange) leaves. Zorvec Enicade is a known oomycide. All leaves were inoculated with mycelial plugs containing *P. agathidicida*. (**d**) Figure C plotting necrosis as a fraction of the leaf over time.

Leafnumber: The ID number of the leaf.

Petriradius: The radius of the Petri dish in pixels.

Petricenter: The coordinates of the center of the Petri dish.

Petriarea: The area of the Petri dish in pixels^2.

Leafarea: The area of the leaf in pixels^2.

Necrosisradius: The distance from start to end of necrotic spread in pixels.

Necrosisarea: The area of the necrosis in pixels.

Plugcenter: The coordinates of the point of infection.

4. Check the images in the *_outlined_images* folder for mistakes. If mistakes are present, run the *manualrepair.py* script. This will prompt the user to first select the folder containing all the images and subsequently select the image that requires fixing. Mistakes should be fixed in the following order: Petri dish, leaves, plugs, and lastly necrosis.
5. To fix Petri dish mistakes, select the image and first double-click in the center of the Petri dish and then on the edge.
6. To fix leaves, double-click on the edge of the leaf and follow the outline of the leaf by repeatedly double-clicking in different points around the leaf. A straight line will be drawn between clicks, so use enough points to relatively accurately trace the edge of the leaf.
7. To fix plugs, double-click in the middle of the plug/point of infection.
8. To fix necrosis, double-click on the edge of necrosis.
9. All of these fixes will be saved when pressing Y, cancelled by pressing N, and skipped by pressing S. This is also indicated in the terminal.
10. The terminal will prompt whether any other fixes on the same image are required. It will then prompt whether any other images are to be fixed.
11. When all classifications for each timepoint are correct, run the *linkleaves.py* script to link leaves from different timepoints (*see* **Note 10**). After running this code, the data from all timepoints

Fig. 4 (continued) (**e**) An example of untreated kauri leaves at the start of an experiment (day = 0). All leaves were inoculated with *P. agathidicida* zoospores (1000 total). (**f**) An example of untreated kauri leaves at the end of an experiment (day = 8). All leaves were inoculated with *P. agathidicida* zoospores (1000 total). (**g**) Example plot of the radius of necrosis (mm) over time in kauri leaves, comparing leaves inoculated with *P. agathidicida* zoospores (1000 total; blue) and uninoculated leaves (negative (-) H_2O control; orange). (**h**) Figure G plotting necrosis as a fraction of the leaf over time. The colored lines in the images (A, B, E, F) delineate the following objects: Petri dish (red), leaf (green), point of infection (blue), and necrosis (black)

per sample are combined, and a plot is generated showing the average development of necrosis in the samples over time (Fig. 4). This development can be expressed in a distance from the point of infection to the edge of necrosis (in pixels or mm), absolute necrosis area (mm^2 or $pixels^2$), or fraction of leaf surface (% of leaf; *see* **Note 11**). This script adds a new column to the *Imageoutput.csv* files, namely:

Filename_original: The filename of the D0 timepoint linked to this image.

Also, a new csv file (*Linkedoutput.csv*) in the main data folder will be created that contains data from all images of all timepoints combined. This output has the same columns as *Imageoutput.csv* files plus two other columns:

Day: The timepoint from start of infection.

Condition: The condition extracted from the file name of the D0 timepoint.

We generate some general plots demonstrating the necrosis development over time, either by area or distance from the time of infection (Fig. 4). This output can also be further analyzed in Python or Excel.

4 Notes

1. It is advisable to perform a preliminary experiment to determine if the necrosis is more visible from the adaxial or abaxial leaf surface and then set up the assay with that side facing upwards for the incision, inoculation, and pictures.
2. In curative assays, the time between inoculation and treatment may vary, depending on the rate of necrotic spread. It may be worth exploring if there is a difference in efficacy when the treatment is applied before vs after visible necrosis.
3. It is easier to deliver the required number of plugs into a sterile vessel (e.g., Petri dish) and use sterilized tweezers to position each plug over the incision.
4. It may be necessary to adjust the inoculum size of 1000 zoospores per leaf, as it is possible that more zoospores may be required to ensure reliable infection or fewer zoospores may be required for differences in resistance to become apparent.
5. For taking photos, any camera can be used, including phone cameras; however, we recommend consistently using the same camera. The relatively small training sets commonly used in these experiments are highly sensitive to changes in cameras. For optimal results, take pictures in a location where lighting conditions can be controlled and remain consistent throughout

the experiment. Try to avoid glare, as it can complicate the training process. Furthermore, the enhancement of necrosis through staining or near-infrared lighting [18] can be used to improve detection sensitivity and accuracy.

6. Although technically these different categories could all be assigned by a single training set, we have found an improvement in performance if different training sets are used for each category. Therefore, we recommend creating a different project for each of these categories.
7. Only a specific subset of your samples is required to be trained, as long as they accurately represent the overall expected variation in the images. We recommend images from four different samples and at timepoints with minimum and maximum necrosis levels.
8. Some pixels may belong to multiple categories simultaneously, such as a pixel that contains necrosis within a leaf inside a Petri dish. Since a separate project is used for each category (*see* **Note 6**), labelling should encompass all pixels specific to that category. For instance, when labelling leaf pixels, both necrosis and plug pixels should be included.
9. Objects are extracted from pixel probability images by searching for patterns specific to each category. For Petri dishes and points of infection (plugs), the algorithm searches for circles. In the case of zoospore inoculation, a circle might not represent the real shape particularly well. However, the quantitative analysis only uses the center of that circle, which should be equivalent between plug- and zoospore-inoculated samples. For necrosis, the algorithm looks for circles that form with a radius extending from the start of the plug and that are bounded within the shape of a leaf. We have found that, for most *Phytophthora* infections in detached leaf assays, this is the best representation of necrosis formation. For leaves, shapes are very flexible, and any shape can be accepted.
10. Linking of leaves can be sensitive to mistakes in leaf segmentation. Ensure that all the leaf objects are correctly delineated. Also, if leaves would have moved significantly during the time course of the experiment, linking might be affected.
11. Different outputs are generated through this analysis pipeline: radius of necrosis (in mm or pixels) or area of necrosis (in mm^2, $pixels^2$, or % of leaf surface). In general, we observed that the radius of necrosis (in mm) has the lowest variability between leaves of the same condition and displays a near-linear trend over the time course of our experiments. However, with small leaves or long experimental time frames, growth can be limited by the dimensions of the leaf; this is most easily detected by plotting the % of leaf area.

References

1. Hernández-Rodríguez A, Miguelez-Sierra Y, Acebo-Guerrero Y, Díaz de la Osa A, Casanova MA (2023) A practical guide to isolation of fluorescent *Pseudomonas* antagonic to *Phytophthora palmivora* (Butler) in *Theobroma cacao*. Physiol Mol Plant Pathol 127:102061. https://doi.org/10.1016/j.pmpp.2023.102061
2. Blossei J, Gäbelein R, Hammann T, Uptmoor R (2022) Late blight resistance in wild potato species – resources for future potato (*Solanum tuberosum*) breeding. Plant Breed 141(3): 314–331. https://doi.org/10.1111/pbr.13023
3. Karki H, Halterman DA (2021) *Phytophthora infestans* (late blight) infection assay in a detached leaf of potato. Bio-Protocol 11(4): e3926. https://doi.org/10.21769/BioProtoc.3926
4. Kliejunas JT (2010) Sudden oak death and *Phytophthora ramorum*: a summary of the literature. https://doi.org/10.2737/PSW-GTR-234
5. Luchi N (2022) Plant pathology: method and protocols. Springer Nature
6. Sarria G, Garcia A, Mestizo Y, Medina C, VarÓN F, Mesa E et al (2021) Antagonistic interactions between *Trichoderma* spp. and *Phytophthora palmivora* (Butler) from oil palm in Colombia. Eur J Plant Pathol 161(4): 751–768. https://doi.org/10.1007/s10658-021-02363-z
7. Shamoun SF, Rioux D, Callan B, James D, Hamelin RC, Bilodeau GJ et al (2018) An overview of Canadian research activities on diseases caused by *Phytophthora ramorum*: results, progress, and challenges. Plant Dis 102(7): 1218–1233. https://doi.org/10.1094/pdis-11-17-1730-fe
8. Sharma BP, Forbes GA, Manandhar HK, Shrestha SM, Thapa RB (2013) Determination of resistance to *Phytophthora infestans* on potato plants in field, laboratory and greenhouse conditions. J Agric Sci 5(5):148–157. https://doi.org/10.5539/jas.v5n5p148
9. Stephan D, Schmitt A, Carvalho SM, Seddon B, Koch E (2005) Evaluation of biocontrol preparations and plant extracts for the control of *Phytophthora infestans* on potato leaves. Eur J Plant Pathol 112(3):235–246. https://doi.org/10.1007/s10658-005-2083-1
10. van den Berg N, Swart V, Backer R, Fick A, Wienk R, Engelbrecht J et al (2021) Advances in understanding defense mechanisms in *Persea americana* against *Phytophthora cinnamomi*. Front Plant Sci 12. https://doi.org/10.3389/fpls.2021.636339
11. Vleeshouwers VGAA, van Dooijeweert W, Paul Keizer LC, Sijpkes L, Govers F, Colon LT (1999) A laboratory assay for *Phytophthora infestans* resistance in various *Solanum* species reflects the field situation. Eur J Plant Pathol 105(3):241–250. https://doi.org/10.1023/A:1008710700363
12. Hodgson WA (1962) Studies on the nature of partial resistance in the potato to *Phytophthora infestans*. Am Potato J 39(1):8–13. https://doi.org/10.1007/BF02912626
13. Annabel LSP, Annapoorani T, Deepalakshmi P (2019) Machine learning for plant leaf disease detection and classification – a review. In: 2019 International conference on communication and signal processing (ICCSP): IEEE, p 0538–42
14. Chakraborty KK, Mukherjee R, Chakroborty C, Bora K (2022) Automated recognition of optical image based potato leaf blight diseases using deep learning. Physiol Mol Plant Pathol 117:101781
15. Sholihati RA, Sulistijono IA, Risnumawan A, Kusumawati E (2020) Potato leaf disease classification using deep learning approach. In: 2020 international electronics symposium (IES): IEEE, p 392–7
16. Sujatha R, Chatterjee JM, Jhanjhi N, Brohi SN (2021) Performance of deep learning vs machine learning in plant leaf disease detection. Microprocess Microsyst 80:103615
17. Berg S, Kutra D, Kroeger T, Straehle CN, Kausler BX, Haubold C et al (2019) Ilastik: interactive machine learning for (bio) image analysis. Nat Methods 16(12):1226–1232
18. Zahid MA, Sandroni M, Vetukuri RR, Andreasson E (2021) A fast, nondestructive method for the detection of disease-related lesions and wounded leaves. BioTechniques 71(2):425–430

Chapter 8

Quantification of *Phytophthora palmivora* Infection in Barley and Related Monocot Roots

Matthew Macleod, Sabine Brumm, and Sebastian Schornack

Abstract

Biotic stresses such as fungal pathogens significantly affect global crop yields. Understanding of the plant-pathogen interactions during root infection, especially in monocot crops, remains limited compared to fungal colonizations of dicots. The infection process of several cereal crop root-damaging fungi and oomycetes is highly similar to root infections by the pathogen model *Phytophthora palmivora*. Here, we describe a protocol for quantifying root infection of *Phytophthora palmivora* in barley or related monocots using reverse transcriptase quantitative PCR. This method can be used to produce relative expression values for the oomycete within roots based on RNA or DNA. Different mutations in the host can, therefore, be quantitatively assessed for their impact on infection. This allows us to identify plant processes underpinning susceptibility to filamentous fungal colonization within monocot models and thus can facilitate the development of possible resistance strategies.

Key words *Phytophthora*, *Phytophthora palmivora*, Barley, *Hordeum vulgare*, Monocot, Infection assay, Oomycete, qPCR

1 Introduction

Pathogens and pests inflict significant global and regional losses to yields in food crop production. The top five most consumed crops globally can experience losses in yield between 15% and 30% and higher losses in region-specific production hotspots [1]. As global climate changes continue to follow predicted trends, an increase in the susceptibility of economically important crops to different plant pathogens is expected [2], as well as the potential for the emergence of new strains of existing pathogens [3]. Among the threats to economically relevant crops, fungal pathogens are an important group and can affect various tissues in the plant. Examples such as rusts (*Puccinia* spp.), head blight (*Fusarium graminearum*), and powdery mildew (*Blumeria graminis)* can impact the monocot crops barley and wheat [1, 4, 5]. While these fungi have been studied infecting plant tissue above the ground, there has been

Monica L. Gerth and Rosie E. Bradshaw (eds.), *Phytophthora: Methods and Protocols*, Methods in Molecular Biology, vol. 2892, https://doi.org/10.1007/978-1-0716-4330-3_8,

less exploration of pathogenic fungi that target monocot roots. Therefore, it is important to develop methods to quantify root infection; such methods could be used for screening potential mutations that may decrease fungal colonization and ultimately improve resistance in crops.

Filamentous pathogenic fungi can be challenging to culture reproducibly and achieving reliable infection of root tissues under laboratory conditions can be difficult. However, the filamentous "fungus-like" oomycete *Phytophthora palmivora* is easy to culture and sporulate and can be genetically transformed with fluorescent markers [5]. *P. palmivora* also reliably infects a broad range of plants, and infection is highly reproducible experimentally using spores, which can easily be set to a standard concentration. Since oomycetes share similar mechanisms to filamentous fungi for invading host cells [6, 7], using *P. palmivora* as a model for this process has a direct contribution to understanding the molecular basis for invasion by filamentous pathogens and possible strategies to reduce it. Our lab and others have studied *P. palmivora* infection in liverworts [8], legumes [9], barley [5], cocoa [10], oil palm [11], and rubber trees [12]. Experimental studies on *P. palmivora* across monocots and dicots have aimed to further characterize its infection mechanism in both leaf and root tissues [9, 13].

Here, we describe a protocol for quantifying infection in *Hordeum vulgare* (barley) (and related monocot) roots by *P. palmivora*. The infection is performed on the roots of seedlings grown on agar plates, and levels of *Phytophthora* RNA are quantified via reverse transcriptase quantitative PCR (RT-qPCR) relative to transcript levels of host housekeeping genes. Normalization of *P. palmivora* transcript levels to a barley housekeeping gene will account for variations in sample preparation and standardize expression for that sample. We use two housekeeping genes for reference to increase the accuracy of this. Barley housekeeping genes were chosen as a baseline, due to their consistent expression levels unaffected by infection. Nucleic acid-based quantification of pathogen biomass can be done directly using extracted DNA, although this captures dead material and may not be sensitive enough to detect low levels of pathogen during early infection. Conversely, reverse transcriptase RNA-based quantification captures the transcriptional profile of the infection, and so markers for highly expressed genes can be used to detect even low levels of pathogen. RT-qPCR starts from a reverse-transcribed DNA copy of total RNA and uses sequence-specific DNA-oligonucleotide primers to amplify a specific region of a target transcript, during which double-stranded DNA of this target accumulates, making this an efficient method for measuring nucleic acid levels with specificity at a time-point [14]. Calculating ratio values for a *P. palmivora* marker transcript relative to housekeeping transcripts allows for the comparison of infection levels between samples or replicates. This relative expression can help visualize the quantitative changes in *P. palmivora* infection across different barley or other monocot varieties and genotypes. Due to the reproducibility of this quantification, it is useful for surveying the germplasm and screening potentially advantageous mutations related to susceptibility to *P. palmivora*.

2 Materials

1. Sterile water.
2. Ethanol (70%).
3. Sodium hypochlorite (bleach).
4. Rotator Mixer (e.g., StarLab).
5. Medium-size plastic plates (e.g., Thermo Scientific™ Nunc™ Lab-Tek Petri Dishes, 79 sq. cm).
6. Whatman paper (e.g., Whatman® 3MM Chr paper 0.34 mm).
7. Forceps.
8. Micropore tape.
9. 24 °C growth cabinet.
10. V8 Vegetable Media For 1 L: 100 mL V8 Juice (V8 brand), 1 g calcium carbonate, 0.05 g β-sitosterol, 15 g Bacto-agar, 900 mL deionized water. Add carbenicillin to eliminate bacterial contamination (25 mg/L).
11. Microbiological safety cabinet as required per local biosafety regulations for work with plant pathogens.
12. Parafilm.
13. Large-size plastic plates (e.g., Thermo Scientific™ 245 × 245 × 25 mm Nunc™ Square BioAssay Dishes).
14. 1% Water Agar (1 g/100 mL agar in water).
15. 15 mL centrifuge tubes.
16. Miracloth (e.g., Merck Millipore Miracloth 22–25 μm pore size).
17. Neubauer Hemocytometer.
18. Transmission Light Microscope with 10× objective.
19. Cellophane discs (325P cellulose, A.A. Packaging Limited).
20. Black bin bag.
21. Stereo/Dissecting Microscope.
22. Flatbed Scanner.
23. 50 mL centrifuge tubes.
24. 2 mL microcentrifuge tubes (e.g., Eppendorf Safe-Lock Tubes).

25. 3 mm tungsten beads for tissue lysis (e.g., Qiagen 3 mm Tungsten Carbide Beads).
26. Liquid Nitrogen.
27. RNA extraction kit (e.g., Sigma-Aldrich Spectrum™ Plant Total RNA Kit).
28. Tissue Lyser (e.g., Qiagen TissueLyser II).
29. On-column DNase kit (e.g., Qiagen RNase-free DNase Set).
30. NanoDrop Spectrophotometer (e.g., Thermo Scientific™ NanoDrop™ One/One^C Microvolume UV-Vis Spectrophotometer).
31. cDNA synthesis kit (e.g., Roche Transcriptor First Strand cDNA Synthesis Kit).
32. Roche LightCycler® 480 SYBR Green I Master Mix.
33. 384-well qPCR plate (e.g., LightCycler® 480 Multiwell Plate 384, white).
34. Clear qPCR plate seal (e.g., LightCycler® 480 Sealing Foil).
35. qPCR machine (e.g., Roche LightCycler® 480).

3 Methods

3.1 Plant Growth and Preparation

This assay tests *Phytophthora palmivora* infection in isolation, and so plants are germinated under sterile conditions to prevent contaminant microorganisms from growing on the nutrient-rich agar medium the plants are moved onto. Sterilization methods will vary depending on the plant system used. Here we will describe a protocol applicable to barley seeds.

1. Collect dried barley seeds in 50 mL centrifuge tubes (~ 40 seeds per tube) and add 40 mL 70% ethanol. Gently invert tubes manually or on a rotating machine (e.g., StarLab Rotator Mixer) for 5 min to ensure all seeds are covered.
2. Pour out the ethanol and wash once with 40 mL of sterile water. Gently invert for 1 min before pouring out.
3. Working in a lateral flow hood, add 40 mL of 4% sodium hypochlorite (bleach) and invert tubes for 10 min, and pour the liquid out. This time can be adjusted to combat any contamination on the seeds (*see* **Note 1**).
4. Perform three further washes with 40 mL of sterile water.
5. Prepare plastic plates (e.g., Thermo Scientific Nunc Lab-Tek Petri Dishes 79 sq. cm or similar) with three pieces of Whatman paper soaked in sterile water (enough so that a small pool forms when the plate is tilted).
6. Use sterile forceps to distribute seeds across the wet paper.

7. Seal plates tightly with Micropore tape and wrap the plates in tinfoil to block light.
8. Place plates at 4 °C for at least 5 d, and then move to 24 °C in a growth cabinet, still in the dark, for 3 d to germinate. *See* **Note 2** for issues with germination.

3.2 Maintenance of P. palmivora Cultures

1. Grow *Phytophthora palmivora* on Petri dishes containing V8 vegetable media at 25 °C in constant darkness, sealed with Parafilm (*see* **Note 3**). *See* **Note 4** for combatting bacterial contamination and for antibiotic selection of *P. palmivora*.
2. To induce sporulation, after 5–7 d growth, place plates under constant light at 24 °C and remove Parafilm to allow the plates to dry for 2–3 d.

3.3 Preparation of Infection Plates

1. Sterilize large plastic plates (e.g., Thermo Fisher 245 × 245 × 25 mm) with 70% ethanol in a microbiological safety cabinet.
2. Pour 1% water agar into these plates (250 mL per plate or until half filled). Ensure that the plates have not dried excessively, as this will hinder plant growth.
3. With sterile forceps, place barley seedlings onto the middle line of the plate from left to right (*see* **Note 5**). It is possible to fit up to 20 seedlings per plate; we have performed assays comparing 5, 8, and 10 replicates of genotypes side-by-side on each half of the plate (*see* **Note 6**).

3.4 Sporulation of P. palmivora for Infection

1. Take the dried *P. palmivora* plates from Subheading 3.2, and flood plates with 5 mL of sterile water for 20–30 min to harvest spores from the cultures (*see* **Note 7**).
2. Transfer the spore solution from the plates into a 15 mL centrifuge tube through sterile Miracloth.
3. Pipette 10 μL of spore solution from the plate onto a Neubauer hemocytometer chamber. Count the spores with a transmission light microscope. Oomycete zoospores should be motile, unlike fungal spores.

Calculate the initial concentration of spores present per mL, and then dilute the spore solution to the desired concentration in sterile water (for more guidance, *see* **Note 8**).

3.5 Root Inoculation with Phytophthora Spores

1. Evenly pipette 100–150 μL of the spore solution (or sterile water as a control, i.e., mock sample) over each individual plant's roots. This volume can be adjusted to achieve the desired total number of spores present on the plate.
2. Place cellophane disks with forceps carefully over roots (around one to two cellophane disks per five seedling roots) to cover

Fig. 1 Arrangement of Barley seedlings on 1% water agar under cellophane discs immediately after inoculation with *Phytophthora palmivora* spores. The *P. palmivora* strain being used is FLIMA (P7545). Three genotypes are being compared (a, b, c) on the same plate with six seedlings per genotype

them and trap *Phytophthora* spore solution (or water) as in Fig. 1. This ensures even distribution of liquid over all parts of each root, as well as between roots of different seedlings (*see* **Note 9**).

3. Seal the edges of the plate with micropore tape.
4. Wrap the bottom of the plate so that the seed and roots are excluded from light (e.g., using aluminium foil or a black bin bag), but the shoots are not, and secure with tape. This simulates the roots being under soil and reduces potential stress and browning induced by light.
5. Leave plates lying flat under constant light at 24 °C. After 1 d, place plates vertically at a 100–110° angle (*see* **Note 10**).

3.6 Tissue Harvesting

1. The number of days allowed for infection depends on the species and strain of *Phytophthora* used and the stage of infection being investigated. Browning symptoms are often visible on the roots at the point of harvesting; however, this merely indicates a strong infection and is not essential. The best method to determine the level of infection visually is under a stereo/dissecting microscope to look for the formation of infection structures.

2. Scan plates with a flatbed scanner to get clear overview images of the infection symptoms.
3. Open plates in a microbiological safety cabinet and discard cellophane disks. Using forceps and a scalpel, cut the root system from each plant and dip it into a 50 mL centrifuge tube filled with sterile water to wash the roots. This way, only plant and oomycete materials are collected and used for PCR reactions. Do the same for any mock samples treated with water.
4. Dry roots with a paper tissue.
5. Place each root system into an individual 2 mL microcentrifuge tube.
6. Add three tungsten beads to each microcentrifuge tube to ensure complete lysis later of the sturdy barley roots.
7. Flash-freeze each tube in liquid nitrogen and store at −80 °C until required for nucleic acid extraction. Wear protective goggles as the liquid nitrogen can pop the tubes open, which could eject the metal beads outwards.

3.7 Nucleic Acid Extraction

Nucleic acid extraction can be performed for RNA or DNA. However, RNA can be a more sensitive marker for infection when there is a lower biomass of *Phytophthora* in the roots, for example, during early infection stages.

1. Carry out RNA extraction using the kit specifications. The Sigma-Aldrich Spectrum ™ Plant Total RNA Kit works effectively (*see* **Note 11**). Ensure that samples are not allowed to thaw (*see* **Note 12**).
2. For the initial preparation, grind the roots in a TissueLyser at 30 Hz for 1 min, then turn the tubes to face the other orientation and grind at 30 Hz for another minute to ensure even grinding of all tissue (*see* **Note 13**).
3. The *Phytophthora* can be considered inactivated completely after the lysis buffer has been added to the ground samples.
4. Perform on column DNase treatment with RNase-Free DNase (e.g., Qiagen), following the manufacturer's instructions.
5. Calculate the concentration of RNA extracted for each sample, for example, with a NanoDrop spectrophotometer.

3.8 cDNA Synthesis

1. Use 1 μg of RNA per sample to synthesize cDNA using a cDNA Synthesis Kit, such as the Roche Transcriptor First Strand cDNA Synthesis Kit, following the manufacturer's recommendations (*see* **Note 14**).

3.9 qPCR

1. Use the following reaction mix: cDNA (2.5 μL) (*see* **Note 15**), SYBR Green (5 μL), Nuclease-free sterile water (0.5 μL), Forward primer (1 μL), Reverse primer (1 μL) (all primers at a concentration of 10 μM). Total volume = 10 μL. Roche LightCycler® 480 SYBR Green I contains a master mix of Taq polymerase and dye.
2. Different primers can be used for measuring the relative *P. palmivora* RNA quantity against Barley housekeeping gene RNA levels.

   Primers that have been used by our group are listed in Table 1, and further primers for marker genes in *P. palmivora* have been described previously [5].
3. Primer efficiencies should be calculated for later use in data analysis (*see* **Note 16**). For each sample, three technical replicates with each primer are used, totalling 12 reactions per plant sample.
4. Pipette master mixes of each reaction into a 384-well qPCR plate, before adding the cDNA and sealing the plate with a clear film. Label the qPCR plate by well, to describe the primer pair, sample and replicate in them (*see* **Note 17**).

   qPCR conditions are as follows: Initial Denaturation 95 °C 5 min, Denaturation 95 °C 20 s, annealing 60 °C 20 s, extension 72 °C 20 s (45× cycles).
5. A melting curve can then be calculated with 1 cycle at 95 °C for 5 s and 65 °C for 1 min, before five continuous acquisitions up to 97 °C (*see* **Note 18**).
6. Sequences of primers that can be used are in Table 1. *Phytophthora palmivora* Elongation Factor 1α (EF1α) (MH760192.1) is a biomass marker, and *H. vulgare* EF1α (KP293846.1) and Cyclophilin (Cyc) (CV056520) are housekeeping genes.

Table 1
Primer sequences for comparison of *Phytophthora palmivora* RNA expression relative to Barley (*Hordeum vulgare*) gene expression

Primer name	Sequence (5′- 3′)
Phytophthora palmivora EF1α forward	5′-CAAGATCCCGTTCGTGCCTA-3′
Phytophthora palmivora EF1α reverse	5′-GCGTTCAGGTTGTCAAGAGC-3′
Hordeum vulgare EF1α forward	5′-ATGATTCCCACCAAGCCCAT-3′
Hordeum vulgare EF1α reverse	5′-ACACCAACAGCCACAGTTTGC-3′
Hordeum vulgare Cyc forward	5′-TTGAGGACGAGATAAGGCCAG-3′
Hordeum vulgare Cyc reverse	5′-GCGACTGACAAGGTGCAAGAG-3′

3.10 Data Interpretation

1. On the LightCycler 480 software, on the analysis page, perform an Absolute Quantification (second Derivative Max) and calculate the values in the table. Export this table as a txt file to create a table of Ct values corresponding to each reaction. Use the $2^{-\Delta\Delta Ct}$ method [15] to calculate the expression of *Phytophthora palmivora* RNA relative to reference housekeeping genes in barley.
2. A boxplot can then be used to effectively visualize relative expression levels across genotypes, treatments, and replicates, and these can be compared using an unpaired Student's T-test.

4 Notes

1. If fungal contamination occurs, for example, under the seed coat of barley seeds, inverting tubes during the bleach step can be extended to 30 min. A longer period of bleaching has proven effective for more rigorous sterilization when contamination is present under the seed coats.
2. If seeds do not germinate evenly in this time, extending the number of days at 4 °C can give more time for germination and synchronization between seeds.
3. Alternating the transfer of *Phytophthora* between spores and plugs week by week ensures a consistent pressure for sporulation within the population. The use of fluorescently tagged *Phytophthora* strains makes imaging more straightforward on plant material [5]. Fluorescence can be checked regularly on subsequent plates to ensure that expression is still strong and that an axenic population is consistently growing.
4. One common issue is bacterial contamination on the plates, which hinders zoospore production and renders cultures unusable for plant infections. To eliminate contamination, carbenicillin (25 mg/L) can be added to V8 media plates. *P. palmivora* can also be transformed with a gentamycin antibiotic-resistance selection marker as described previously [16].
5. Drag seedlings slightly up to ensure roots are in contact with agar and are straightened and pointing to the bottom of the plate (Fig. 1). This ensures maximal contact with the *Phytophthora* solution once it is added over the roots and distributed over the agar. Arranging roots like this will also help them grow evenly downwards and in constant contact with the media. If the roots appear to be shrivelled after a few days on the plates even with a mock treatment, ensure that they are in contact with the agar.
6. Comparisons between infections in different plant genotypes should be performed on the same plates to ensure that growth conditions are identical.

7. Placing *Phytophthora palmivora* plates at 4 °C for 20–30 min prior to flooding with water may also aid in stressing them to stimulate sporulation. It is advisable to stimulate sporulation in additional spare plates of *Phytophthora* in case sporulation is sporadic or poor.
8. Adjust the spore concentration (spores/mL) based on the host plant being affected, the *Phytophthora* species and strain being used, and the stage of infection being investigated. We use *Phytophthora palmivora* accession FLIMA (P7545) and find a concentration of 50,000 spore/mL is suitable when studying earlier stages of *Phytophthora* infection (3 dpi). Our spore concentration is calculated by counting the average number of spores in large-size hemocytometer squares (×10,000 for spores/mL). We advise against an excessively large spore concentration if comparing resistance between genotypes. The high concentration results in high expression of pathogen RNA across all plants, so small differences in resistance between plant genotypes may not be resolved.
9. Air bubbles can often form under the cellophane, so it is useful to carefully push these outwards using forceps, so that the spore solution or water gets evenly distributed around the roots and all plant samples are covered.
10. Leaving plates flat for the first day helps prevent the spore solution from running immediately down the plate and allows the spores to collect and set into the roots. Plates are then grown at a 100–110° angle which prevents the roots from growing into the agar. This makes harvesting an easier process without the need to remove additional agar residue.
11. With the Spectrum RNA kit, protocol A is used in **step 4** for RNA binding as recommended for root tissue. Only a single final elution step is carried out.
12. Keep at −80 °C or in liquid nitrogen until the lysis solution has been added. Dry ice can also be used to temporarily hold samples. For optimal RNA yields, minimize the time taken for grinding tissue and immediately return samples to liquid nitrogen between grinding and the addition of the lysis solution to prevent thawing of the tissue and degradation of RNA.
13. Tissue grinding is recommended to be performed with a grinding pestle (e.g., a StarLab microtube pestle) according to the Sigma Spectrum RNA Kit specifications. However, due to the strength of barley roots, we recommend using a TissueLyser with tungsten beads to give a more rigorous grinding. Leaving samples in the TissueLyser for slightly longer than is recommended helps to guarantee all samples are equally ground to a

fine powder with barley roots. Where some tissue samples may not be completely turned into powder, additional grinding can be done with a pestle. A poor RNA yield can be the result of incomplete grinding of the root tissue.

14. Synthesize cDNA following kit recommendations and using Procedure A with anchored-oligo$(dT)_{18}$ primers. Carry out the optional initial denaturation step.
15. We dilute the cDNA tenfold before the qPCR reactions with our primers.
16. The efficiency of two primer pairs in a reaction should be between 90% and 110%. Primer efficiency can be calculated by performing a serial dilution of cDNA containing the target gene, starting from a concentration of 1 (e.g., 10 ng DNA) and incremental dilutions by a factor of 10 or 2 [17]. These reactions should also be performed in triplicate. Plot the average Ct values for each dilution against the log10 values for each concentration in the serial dilution. Primer efficiency is calculated using the eq. $E = 10^{-(1/slope)}$ ($\times 100$ for %) as done previously [17, 18].
17. We recommend labelling each well of the qPCR plate with the sample and primer combination on the LightCycler software. This will save time later when matching the Ct values in the qPCR output file to the reaction they correspond to.
18. Melt curve analysis shows whether binding has occurred for a primer pair to a secondary amplicon during the reaction. A melt curve analysis should be done both with primer efficiency calculation and during sample testing to confirm the specificity of primer-target binding. A single bell-shaped curve indicates that the primer pair has only bound to the target gene. However, a second, smaller bell-shaped curve could indicate that primer dimerization or non-target binding has occurred, and the reaction may need to be repeated in case contamination has occurred in one of the reagents. Mock-treated (water) root samples should not have any amplification by the *P. palmivora*-specific primers, and so the melt curve for these reactions will not have any peaks.

Acknowledgements

This work was funded by UKRI grant BB/X014118/1 to S.S.

References

1. Savary S, Willocquet L, Pethybridge SJ et al (2019) The global burden of pathogens and pests on major food crops. Nat Ecol Evol 3: 430–439. https://doi.org/10.1038/s41559-018-0793-y
2. Eckardt NA, Ainsworth EA, Bahuguna RN et al (2023) Climate change challenges, plant science solutions. Plant Cell 35:24–66. https://doi.org/10.1093/plcell/koac303
3. Velásquez AC, Castroverde CDM, He SY (2018) Plant–pathogen warfare under changing climate conditions. Curr Biol 28:R619–R634. https://doi.org/10.1016/j.cub.2018.03.054
4. Dean R, Van Kan JAL, Pretorius ZA et al (2012) The top 10 fungal pathogens in molecular plant pathology. Mol Plant Pathol 13:414–430. https://doi.org/10.1111/j.1364-3703.2011.00783.x
5. Le Fevre R, O'Boyle B, Moscou MJ, Schornack S (2016) Colonization of barley by the broad-host hemibiotrophic pathogen *phytophthora palmivora* uncovers a leaf development–dependent involvement of mlo. Mol Plant-Microbe Interactions® 29:385–395. https://doi.org/10.1094/MPMI-12-15-0276-R
6. Bozkurt TO, Kamoun S (2020) The plant–pathogen haustorial interface at a glance. J Cell Sci 133:jcs237958. https://doi.org/10.1242/jcs.237958
7. Panstruga R, Dodds PN (2009) Terrific protein traffic: the mystery of effector protein delivery by filamentous plant pathogens. Science 324:748–750. https://doi.org/10.1126/science.1171652
8. Carella P, Gogleva A, Tomaselli M et al (2018) *Phytophthora palmivora* establishes tissue-specific intracellular infection structures in the earliest divergent land plant lineage. Proc Natl Acad Sci 115:E3846–E3855. https://doi.org/10.1073/pnas.1717900115
9. Rey T, Chatterjee A, Buttay M et al (2015) Medicago truncatula symbiosis mutants affected in the interaction with a biotrophic root pathogen. New Phytol 206:497–500. https://doi.org/10.1111/nph.13233
10. Baruah IK, Ali SS, Shao J et al (2021) Changes in gene expression in leaves of cacao genotypes resistant and susceptible to *Phytophthora palmivora* infection. Front Plant Sci 12:780805. https://doi.org/10.3389/fpls.2021.780805
11. Avila-Mendez K, Rodrigo Á, Araque L, Romero HM (2019) Simultaneous transcriptome analysis of oil palm clones and *Phytophthora palmivora* reveals oil palm defense strategies. PLoS One 14:e0222774. https://doi.org/10.1371/journal.pone.0222774
12. Deenamo N, Kuyyogsuy A, Khompatara K et al (2018) Salicylic acid induces resistance in rubber tree against *Phytophthora palmivora*. Int J Mol Sci 19:1883. https://doi.org/10.3390/ijms19071883
13. Evangelisti E, Guyon A, Shenhav L, Schornack S (2023) FIRE mimics a 14-3-3-binding motif to promote *Phytophthora palmivora* infection. Mol Plant-Microbe Interactions® 36:315. https://doi.org/10.1094/MPMI-12-22-0251-R
14. Yuan JS, Reed A, Chen F, Stewart CN (2006) Statistical analysis of real-time PCR data. BMC Bioinformatics 7:85. https://doi.org/10.1186/1471-2105-7-85
15. Livak KJ, Schmittgen TD (2001) Analysis of relative gene expression data using real-time quantitative PCR and the 2−ΔΔCT method. Methods 25:402–408. https://doi.org/10.1006/meth.2001.1262
16. Evangelisti E, Yunusov T, Shenhav L, Schornack S (2019) *N*-acetyltransferase AAC(3)-I confers gentamicin resistance to *Phytophthora palmivora* and *Phytophthora infestans*. BMC Microbiol 19:265. https://doi.org/10.1186/s12866-019-1642-0
17. Jozefczuk J, Adjaye J (2011) Chapter six – quantitative real-time PCR-based analysis of gene expression. In: Jameson D, Verma M, Westerhoff HV (eds) Methods in enzymology. Academic Press, pp 99–109. https://doi.org/10.1016/B978-0-12-385118-5.00006-2
18. Rasmussen R (2001) Quantification on the LightCycler. In: Meuer S, Wittwer C, Nakagawara K-I (eds) Rapid cycle real-time PCR: methods and applications. Springer, Berlin, Heidelberg, pp 21–34. https://doi.org/10.1007/978-3-642-59524-0_3

Chapter 9

The Use of eDNA Metabarcoding to Detect and Identify *Phytophthora* in Water Samples

Eva Randall, Beatrix Keillor, and David E. L. Cooke

Abstract

We describe a protocol to amplify DNA barcodes of known and unknown taxa of *Phytophthora* and related plant pathogenic oomycetes from a range of environments. The methods focus on sampling pathogen propagules from water using in situ sampling and filtration equipment and buffers that enable efficient storage and DNA extraction for later downstream processing.

Key words Environmental DNA, ITS rDNA, Filtration, Sampling

1 Introduction

Environmental DNA (eDNA) metabarcoding is a powerful method of assessing biodiversity whereby samples of water, soil, sediment, or air are processed by extracting DNA from which DNA barcodes are amplified using general or universal primers in a polymerase chain reaction (PCR). The DNA barcodes are then sequenced using next-generation sequencing (NGS) to generate thousands to millions of reads that are analysed to determine the species composition in the sample [1]. Despite the rapid expansion of this method, limitations and challenges in the experimental procedures and processes remain, including the need for robust sampling procedures, effective sample preservation methods and means of minimizing the risks of contamination during sampling, DNA extractions, and other downstream processing.

Water is a commonly sampled substrate for *Phytophthora* eDNA metabarcoding, but the preservation of water eDNA can be a challenge, particularly in warmer environments [1]. Preventing sample degradation often relies upon low-temperature storage during field sampling, transport and in the laboratory. This poses practical limitations for field collections in locations where refrigeration and frozen storage are difficult or where samples must be

Monica L. Gerth and Rosie E. Bradshaw (eds.), *Phytophthora: Methods and Protocols*, Methods in Molecular Biology, vol. 2892, https://doi.org/10.1007/978-1-0716-4330-3_9,

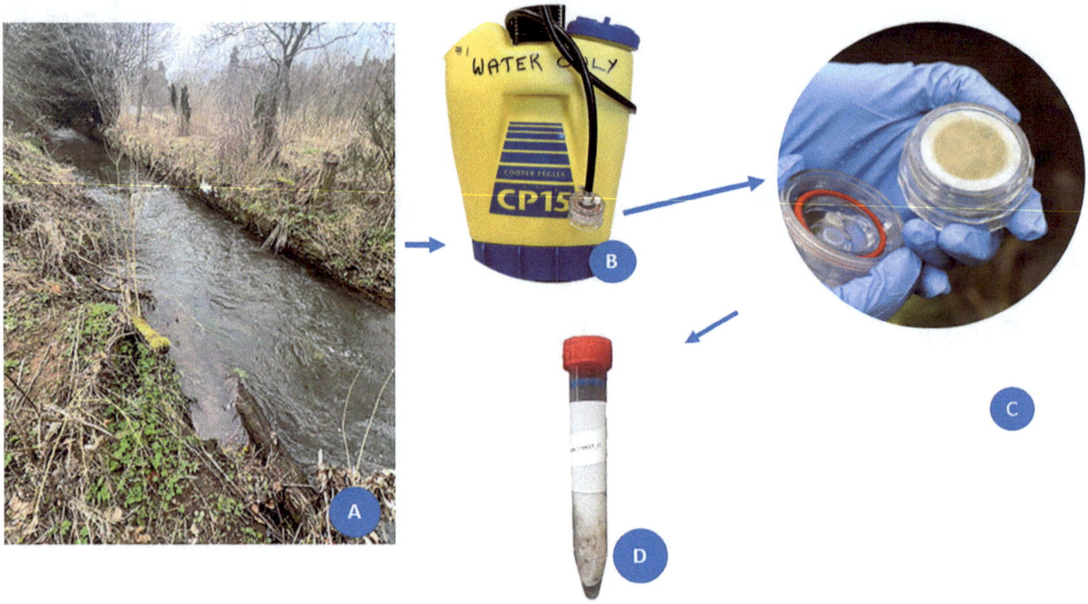

Fig. 1 Sampling area and tools for *Phytophthora* propagule capture. (**a**) Sampling area; (**b**) Knapsack with polypropylene filter holder; (**c**) Filter holder open with used filter; (**d**) Labelled 15 mL sterile tube containing Longmire preservation buffer and filter

transported long distances for further processing and screening [1, 2]. To minimize eDNA degradation, preservatives such as ethanol, Longmire lysis/preservation buffer (from here on called Longmire preservation buffer), benzalkonium chloride (10% w/v), and RNAlater have been used in combination with low temperatures and freezing [2–5]. Recently, kits for eDNA sampling such as the NucleoSpin® eDNA water (Macherey-Nagel, Germany), Complete eDNA sampling set (Sylphium Molecular Ecology, Netherlands), Filter eDNA Collection Kit (SureScreen Scientifics Ltd, England) have become available for eDNA water sampling to minimize the risks of cross-contamination and allow storage at room temperature.

The filtration of water samples from aquatic environments to capture propagules of *Phytophthora* (zoospores, cysts germinating cysts, etc.) isolates is becoming a routine eDNA metabarcoding technique (Fig. 1) [6–12]. In our laboratory in Dundee, this technique was first tested in 2012 in Scibetta's work to assess the *Phytophthora* diversity in samples of soil and water [6]. As the work expanded with multiple in situ field-filtered water samples, Longmire preservation buffer in combination with low-temperature storage to preserve eDNA samples during transportation was tested with very good results. However, with increasing numbers of samples, it became clear that an alternative to the laborious and hazardous phenol-chloroform-isoamyl alcohol eDNA extraction method was required.

Several kits have been reported in the literature for aquatic eDNA extraction: e.g., Qiagen DNeasy Blood and Tissue kit and DNeasy Power Soil kit (Qiagen, Hilden, Germany); 1-step DNA Isolation Kit for Tissues & Cells (nexttec Biotechnologie GmbH, Germany); E.Z.N.A Plant Kit (Omega Bio-tek, Doraville, USA) [3, 5, 13]. We opted for the Qiagen DNeasy Blood and Tissue kit for eDNA extractions as it had been used successfully for aquatic eDNA extraction with minor changes to the manufacturer's procedures [2, 3, 13].

For eDNA PCR screening of *Phytophthora* propagules, our assay uses a semi-nested ITS1 gene region PCR protocol with Scibetta's *Phytophthora*-specific primers [6] to amplify an approximately 262 bp barcode. The test amplifies all known clades of *Phytophthora*, *Nothophytophthora,* and the known genera of downy mildews. Since the proportion of *Phytophthora* propagules in most eDNA water samples is low, we find two rounds of PCR are required. The semi-nested PCR approach increases sensitivity but also increases the risk of cross-contamination [8] and amplification of false positives, which can be minimized by carefully following this protocol and the use of synthetic DNA barcode controls. The synthetic controls are four distinct randomly generated sequences of an average length and base composition for *Phytophthora* ITS1 regions.

In this chapter, we describe the water sampling and storage procedures, eDNA extraction, PCR sample screening, barcode PCR, and sample clean-up for Illumina MiSeq sequencing of an rDNA ITS1 barcoding assay for *Phytophthora* and closely related taxa (Figs. 1 and 2).

2 Materials

Prepare a data sheet with an identification (ID) code and date for each sampling site and standardized metadata (*see* **Note 1**). Prepare all solutions and buffers using purified deionized autoclaved water and analytical grade reagents and store at room temperature (unless indicated otherwise). Prepare fresh solutions prior to use.

2.1 Buffer and Field Equipment/Tools

1. Longmire preservation buffer: 0.1 M Tris–HCl, 0.1 M EDTA, 10 mM NaCl, 0.5% (w/v) SDS, pH 8.0 (*see* **Note 2**).
2. Chlorine tablet (Palintest Instachlor tablets PR-150, code 424) (*see* **Note 3**).
3. Knapsack sprayer (CP15 2000 Series 15 L) (*see* **Note 4**).
4. Polypropylene filter holder (PALL 47 mm in-line; Part No. 1119 (*see* **Note 4**).

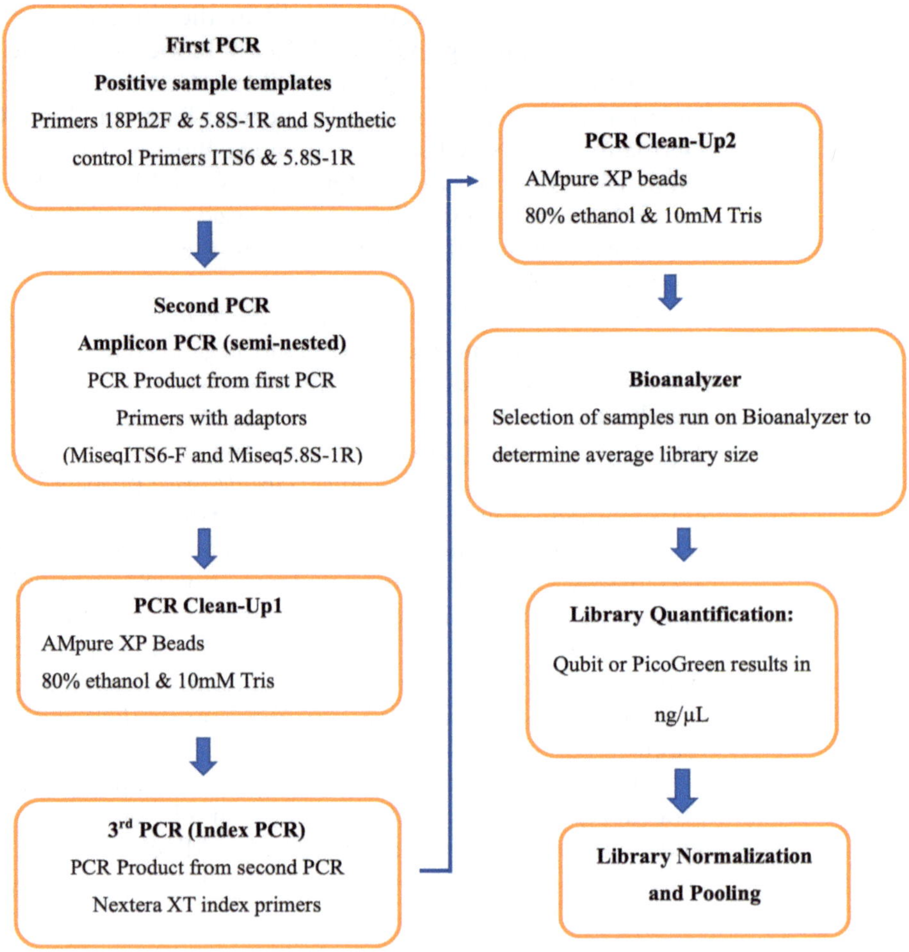

Fig. 2 *Phytophthora* ITS1 metabarcoding library preparation workflow

5. Cellulose ester filters, 47 mm diameter, 1.2 μm pore size, polyester web, Millipore (RW1904700).
6. Labelled 15 mL sterile tubes containing 8 mL of Longmire preservation buffer (*see* **Note 5**).
7. General equipment: Sheets of pre-filter Muslin cloth (*see* **Note 6**), forceps and scissors, a bucket, nitrile gloves, plastic bags for waste, and a cool box with pre-frozen ice packs.

2.2 Lab Equipment and Tools

1. 2 mL Eppendorf tubes.
2. Tube rotator.
3. Heating block.
4. PCR cabinet with UV light.
5. PCR machine.
6. 96-well plate centrifuge and vortex plate adapter.
7. 10% Bleach (0.45% Sodium hypochlorite) (*see* **Note 7**).

2.3 DNA Extraction and PCR

1. PBS buffer: 137 mM NaCl, 2.7 mM KCl, 8 mM $Na_2HPO_4 \cdot 12H_2O$, and 2 mM KH_2PO_4, pH 7.4.
2. Qiagen DNeasy Blood and Tissue Kit.
3. PCR-grade water.
4. Proteinase K.
5. PCR primers (Tables 1 and 2).
6. Synthetic DNA controls (268 bp long) with start and end sequences identical to round 2 PCR primers (Integrated DNA Technologies). See GenBank Accession numbers PP407413 to PP407416.
7. KAPA HiFi HotStart ReadyMix (2×) (Roche Sequencing, *see* **Note 8**).

2.4 PCR Product Clean-Up and Library Preparation

1. 10 mM Tris–HCl pH 8.5 (*see* **Note 9**).
2. 80% ethanol (*see* **Note 10**).
3. Agencourt AMPure XP 60 mL kit (Beckman Coulter Genomics) (*see* **Note 11**).
4. 96-well PCR plate, 0.2 mL.
5. Adhesive PCR plate sealing sheet.
6. 7 mL sterile bijou container (e.g., Sterilin, Medline Scientific, Runlab).
7. Sterile reagent reservoir 25 mL for multichannel pipette.
8. 50 mL vials, timer, ice bucket, and clean lab coats.
9. DynaMag 96-well plate magnetic stand.
10. 1000, 200, 20, and 10 μL manual single-channel pipettes (*see* **Note 12**).
11. 200, 20, and 10 μL manual multichannel pipettes with volume lock.

Table 1
PCR primers used for first sample screen against *Phytophthora* and related oomycetes

Direction	Primer name	Primer sequence (5′ – > 3′)[a]	Amplicon size
First round PCR			
Forward	18Ph2F	GGATAGACTGTTGCAATTTTCAGT	~469 bp
Reverse	5.8S-1R	GCARRGACTTTCGTCCCYRC	
Second round PCR			
Forward	ITS6F	GAAGGTGAAGTCGTAACAA	~262 bp
Reverse	5.8S-1R	GCARRGACTTTCGTCCCYRC	

[a]Note IUPAC ambiguity codes used for some bases

Table 2
PCR primers used for *Phytophthora* barcoding PCRs (*see* Note 22)

Direction	Primer name	Primer sequence (5′ – > 3′)	Amplicon size
First PCR (as Table 1)			
Second PCR (semi-nested)			
Forward	MIS_ITS6_F	**TCGTCGGCAGCGTCAGATGTGTATAAGAGACAG**GAAGGTGAAGTCGTAACAAGG[a]	~400 bp
Reverse	MIS_5.8S-1R	**GTCTCGTGGGCTCGGAGATGTGTATAAGAGACAG**GCARRGACTTTCGTCCCYRC[a]	

[a]The bold underlined letters in the primer sequences represent Illumina overhang adapters

12. 200, 20, and 10 μL filter pipette tips.
13. 1 L and 200 mL autoclaved bottles and 1 L, 250- and 100 mL measuring cylinders to be used only for preparing ethanol (*see* **Note 12**).
14. Nextera XT Index Kit v2 Set A (96 indexes, 384 samples) (FC-131–2001) and Set D (96 indexes, 384 samples) (FC-131–2001) (FC-131-2004).
15. 5 mL Eppendorf tube required to accommodate total volume, if preparing a plate with 96 samples.
16. Quant-iT PicoGreen assay (ThermoFisher Scientific).
17. Standard fluorescent microplate reader.
18. Qubit fluorimeter.
19. Qubit 1× dsDNA HS Assay Kit (0.1–120 ng).
20. Bioanalyzer DNA 2100.

3 Methods

The method was developed to sample *Phytophthora* propagules from water bodies such as streams, lakes, and puddles, or water collected after passing through compost of commercially produced plants in nurseries. All laboratory procedures are carried out on the bench at room temperature, unless specified.

3.1 Pre-sampling Stage

1. If more than one knapsack sprayer is used for field sampling, label the knapsacks.
2. Place a filter in the filter holder and pass 5 L of mains tap water through each knapsack as a laboratory control.
3. Place the filter in the 15 mL vial with Longmire preservation buffer.
4. Repeat this step twice to generate three blank filters for each knapsack.
5. Store all three blank filters from the same knapsack in the same vial.
6. Label vials and store at −20 °C.

3.2 Field Sampling from Streams and Water Sources at Nurseries

1. At each sampling point, collect 5 L of water, and pour through the muslin cloth (if required) into the knapsack sprayer (*see* **Note 13**).
2. Place a filter in the filter holder and, using the knapsack to generate sufficient pressure, filter the water (*see* **Note 14**).

3. When the filter begins to block, remove and submerge in Longmire preservation buffer in the 15 mL sampling vial (*see* **Note 15**).
4. Repeat this step until three filters are blocked or all 5 L of water has been filtered.
5. Filters from the same 5 L sample are stored in a single 15 mL vial.
6. Repeat **steps 1–5** twice to generate three replicates per sampling point. Between each sampling point, clean the knapsack and buckets with approximately 3 L of water and a chlorine tablet left for approximately 30 min to dissolve. Flush sprayer pipes and, after soaking, rinse with clean water to remove residual chlorine (*see* **Note 16**).
7. Store vials with filters and buffer in a cool box during sampling (*see* **Note 15**).
8. All solid waste produced is kept and autoclaved in the lab.
9. All equipment should be thoroughly cleaned in the laboratory between sampling trips.

3.3 Laboratory Sample Preparation

1. Clean the bench with 10% bleach before and after handling samples (*see* **Note 17**).
2. Check filters are submerged in preservation buffer. Top up if necessary and store at 5 °C (*see* **Note 18**).
3. Place each tube on a rotating wheel mixer for 10 min at room temperature.
4. Remove filters from the preservation buffer and place into a 2 mL Eppendorf and store at −20 °C as a backup.
5. Transfer a 1.5 mL aliquot of each sample in Longmire preservation buffer to a 2 mL Eppendorf for DNA extraction.
6. Store aliquot and remaining 15 mL sample preservation buffer at −20 °C.
7. Repeat **steps 2–5** with blank control filters stored at −20 °C.

3.4 DNA Extraction from Preservation Buffer Containing eDNA

The procedures for eDNA extraction from samples in preservation buffer follow the protocols described in the Qiagen DNeasy Blood & Tissue DNA (Spin-Column Protocol) [14] with minor changes to accommodate the different substrate (*see* **Note 19**).

1. Prior to eDNA extraction, fully defrost the aliquots of samples in preservation buffer.
2. Centrifuge the sample preservation buffer aliquots for 5 min at 3000× *g*.
3. Remove and dispose of the supernatant.
4. Resuspend the remaining pellet in 200 μL 1× phosphate-buffered saline (PBS) and add 20 μL proteinase K.

5. Add 200 μL Buffer AL (lysis buffer), mix by vortexing and incubate samples at 56 °C for 10 min.
6. Add 200 μL ethanol (96–100%) and mix by vortexing.
7. Pipette the mixture into a DNeasy mini spin column placed in a 2 mL collection tube. Centrifuge at ≥6000× *g* for 1 min. Discard the flow-through and collection tube.
8. Place the spin column in a new 2 mL collection tube, add 500 μL Buffer AW1 (wash buffer 1) and centrifuge for 1 min at ≥6000× *g*. Discard the flow-through and collection tube.
9. Place the spin column in a new 2 mL collection tube and add 500 μL Buffer AW2 (wash buffer 2). Centrifuge for 3 min at 20,000× *g*. Discard the flow-through and collection tube.
10. Transfer the spin column to a new 1.5 or 2 mL microcentrifuge tube.
11. Elute the DNA by adding 50 μL elution Buffer AE to the center of the spin column membrane. Incubate for 1 min at room temperature (15–25 °C). Centrifuge for 1 min at ≥6000× *g* and store the eluted sample DNA at −20 °C (*see* **Note 19**).

3.5 eDNA Screen PCR

The reactions and conditions for both rounds of PCR are identical (Tables 3 and 5) except for the different forward primer in the second round (Table 1) and that 1 μL of the first round PCR product is used as a DNA template for the second round. All PCRs are prepared inside a PCR cabinet, and both PCR master mixes are prepared prior to handling any eDNA to minimize the risks of contamination. This screen of all samples is to identify those

Table 3
First round PCR and second round PCR (semi-nested) reactions to screen eDNA samples and first round PCR for synthetic controls

eDNA samples		Synthetic controls	
Reagents	**Volume (1×)**	**Reagents**	**Volume (1×)**
PCR grade water	4.51 μL	PCR grade water	0.51 μL
2× KAPA HiFi HotStart ReadyMix	6.25	2× KAPA HiFi HotStart ReadyMix	6.25
10 μM Primer F	0.37	10 μM Primer F (ITS6F)	0.37
10 μM Primer R	0.37	10 μM Primer R (5.8S_1R)	0.37
Template DNA ng/μL	1	Template DNA pg/μL[b]	5
Volume	12.5 μL[a]	Volume	12.5 μL[c]

[a]Dispense 11.5 μL in wells
[b]Picogram per microliter
[c]Dispense 12.5 μL in the designated wells

to be taken forward for barcoding (*see* **Note 20**). Amplicons are visualized by electrophoresis on 2% agarose gels, and positive samples are separated and stored at −20 °C.

1. Prepare the 96-well plate layout using previously defined Sample IDs (*see* **Note 1**).
2. Prepare the working area by cleaning the bench and pipettes with 10% bleach. Wipe clean the PCR cabinet with a damp cloth and place all the items needed for PCR except DNA and reagents into the cabinet and UV treat for 15 min.
3. Defrost, vortex, and centrifuge reagents.
4. Prepare master mix reactions for both PCRs as described in Tables 3 an 4.
5. Label and date the 96-well plates for the first and second round PCR and dispense 11.5 μL of the master mix accordingly onto each plate and seal.
6. Centrifuge plates with master mix at 1000× *g* for 1 min.
7. Store PCR second round plate on ice.
8. Defrost, vortex, and centrifuge eDNA samples.
9. Add 1 μL of DNA into each well of the first round PCR plate and gently pipette the entire volume up and down ten times to mix.
10. Seal plate and centrifuge at 1000× *g* for 1 min.
11. Run the first round PCR with the conditions described in Table 5.
12. Proceed to the second round PCR.
13. Centrifuge PCR first round plate at 1000× *g* for 1 min.
14. Carefully remove the seals from both PCR plates.
15. Transfer 1 μL of PCR product from the first round PCR into each well of the second round plate and gently mix entire volume.
16. Seal plate and centrifuge at 1000× *g* for 1 min.
17. Run second round PCR (semi-nested) with the conditions described in Table 5.
18. Run 5 μL of PCR product on a 2% agarose gel.
19. Samples with a PCR product of the expected size (Table 1) are identified, and the corresponding eDNA samples are separated and stored at −20 °C for barcoding.
20. Discard PCR plates.

3.6 Amplicon PCR

Positive samples from Subheading 3.5 are used to prepare ITS1 metabarcoding libraries (*see* **Notes 21** and **22**) and the synthetic DNA controls are added in this step. A pre-prepared mix of four synthetic DNA fragments diluted to a range of concentrations (10 ag (attogram), 100 ag, 1000 ag, and 10 fg (femtogram) per reaction) are used as controls for the ITS1 metabarcoding libraries. Add the synthetic control mix to six samples per 96-well plate, spread across the zones of the plate in wells B02, F03, B06, F07, B10, and F11.

Two PCRs are carried out: First round PCR for positive samples with the same set of primers and reaction volume previously described (Table 3) and for synthetic controls with ITS6F and 5.8S-1R primers (Table 3, *see* **Note 23**). For the second round PCR (semi-nested) with the Illumina overhang adapter sequence primers (Table 2), 1 μL from the first round PCR product is used as the DNA template (Table 5). Keep the first PCR product plate and store at −20 °C.

1. Repeat PCRs as per **steps 1–17** from Subheading 3.5—with the following exceptions:
   (a) The second-round reaction must follow conditions in Table 4 in which the reaction volume is increased to 30 μL. Therefore in **Step 5**, 29 μL of master mix needs to be dispensed into the second-round plate.
   (b) Second-round PCR conditions have a different final extension time (Table 6).
2. Transfer 25 μL of the second round PCR product to a new amplicon plate and proceed to the PCR clean-up step.
3. Store the second round PCR product plate with the remaining 5 μL of reaction mix in the fridge to run on a 2% agarose gel (*see* **Note 24**).

Table 4
Second PCR reaction with Illumina overhang adaptor primers (semi-nested)

Reagents	1×	Final conc.
PCR water grade	12.2	
2× KAPA HiFi HotStart ReadyMix [15]	15	
10 μM Primer F (Mis_ITS6F)	0.9	0.3 μM
10 μM Primer R (Mis_5.8S_1R)	0.9	0.3 μM
Template DNA	1 μL	
Volume[a]	30 μL[b]	

[a]Dispense 29 μL of the master mix reaction into each well
[b]Total volume based on 30 μL of amplicon, 4 μL to run on a gel, 1 μL to account for error

Table 5
PCR conditions for screen with *Phytophthora* primers (*see* Note 25)

Steps	Temperature	Time
Initial activation	95 °C	3 min
Step cycling	25 cycles	
Denaturation	98 °C	20 s
Extension	61 °C	25 s
Annealing	72 °C	20 s
Final extension	72 °C	1 min
Hold	4 °C	∞

Table 6
PCR conditions for semi-nested second round and for primers with Illumina overhang adaptors

Steps	Temperature	Time
Initial activation	95 °C	3 min
Step cycling	25 cycles	
Denaturation	98 °C	20 s
Extension	61 °C	25 s
Annealing	72 °C	20 s
Final extension	72 °C	5 min
Hold	4 °C	∞

3.7 Amplicon PCR Clean-Up 1 (Protocol Format for 96-Well and 25 μL PCR)

This involves binding the amplicon DNA to magnetic beads followed by two wash and elution steps to efficiently remove contaminants such as dNTPs, primers, and primer-dimer products. The procedures for Amplicon PCR clean-up follow principally the Illumina protocol [16] with minor changes (*see* **Notes 25** and **26**). All consumables and quantities required for this step are summarized in Table 7.

1. Defrost a 50 mL aliquot of 10 mM Tris–HCl pH 8.5 and, 1 h prior to starting the procedure, remove a 2.5 mL AMPure XP bead working aliquot from the fridge and leave to reach room temperature (*see* **Note 11**).

Table 7
PCR clean-up 1 consumables and quantities

Name	Quantities	Storage
AMPure XP beads [17]	20 μL per sample	Aliquots of 2.5 and 3.0 mL prepared and stored in the fridge at 2° to 8 °C
80% ethanol	400 μL per sample	200 mL freshly prepared
10 mM Tris pH 8.5	52.5 μL per sample	Aliquots of 50 mL prepared and stored between −15° and −25 °C. Aliquots can be stored up to 6 months
PCR reservoir for multichannel 25 mL	3	
PCR seal sheets	3	
96-well plate	2 plates	

2. Prepare 80% ethanol (*see* **Note 10**).
3. Centrifuge the amplicon PCR plate at 1000× *g* at 20 °C for 1 min and carefully remove the PCR plate seal.
4. Vortex the aliquot of AMPure XP beads for 30 s and pour into a sterile reservoir (*see* **Notes 27** and **28**).
5. Using a multichannel pipette, carefully add 20 μL of AMPure XP beads to one column of the amplicon PCR plate and gently pipette entire volume up and down ten times. Change tips between columns and repeat this step as required (*see* **Note 29**). Discard beads reservoir.
6. Incubate at room temperature without shaking for 5 min.
7. Place the amplicon PCR plate on the magnetic stand for 2 min or until the supernatant has cleared (*see* **Note 30**).
8. Keep the amplicon PCR plate on the magnetic stand and with a 200 μL multichannel pipette set at 45 μL carefully remove and discard the supernatant. Change tips between columns (*see* **Note 31**).
9. Two washes are performed between **steps 10** and **13**. All steps must be carried out with the amplicon PCR plate on the magnetic stand (*see* **Note 32**).
10. First wash: Add 200 μL of freshly prepared 80% ethanol into each sample well. Change tips between columns (*see* **Note 33**).
11. Incubate sample plate on the magnetic stand for 30 s.
12. Using a 200 μL multichannel pipette gently remove ethanol and discard. Change tips between columns (*see* **Notes 31** and **34**).

13. Second wash: Repeat **steps 10–12**, then continue to **step 14**.
14. Check if all ethanol has been removed (*see* **Note 35**) and note that any remaining ethanol can be removed with smaller volume multichannel or single pipettes.
15. Leave the amplicon PCR plate on the magnetic stand, to allow the beads to air-dry for 10 min (*see* **Note 36**).
16. Remove the sample plate from the magnetic stand. Pour 25 mL of 10 mM Tris–HCL pH 8.5 (elution buffer) into a new reservoir and add 52.5 μL of the elution buffer to each well ensuring pellets are covered (*see* **Note 37**).
17. Gently mix the entire volume up and down ten times until all beads are fully resuspended, changing tips between columns (*see* **Notes 29** and **38**).
18. Incubate at room temperature for 2 min.
19. Place the amplicon plate on the magnetic stand for 2 min until the supernatant has cleared (*see* **Note 30**).
20. Carefully transfer 50 μL of the eluate to a new 96-well PCR plate, changing tips between columns (*see* **Note 39**).
21. Seal amplicon PCR plate and centrifuge at 1000× *g* for 1 min.
22. Carefully remove the seal and transfer 5 μL to a new PCR plate to run libraries on a 2% agarose gel.
23. Seal amplicon PCR plate. Store at −20 °C for up to 1 week (safe stopping point) as recommended in the Illumina protocol.
24. Run the 4 μL and 5 μL PCR products from before and after the clean-up step on an agarose gel and compare the results (*see* **Note 40**). The expected amplicon size is ~400 bp.

    The PCR amplicon plate product contains the purified DNA barcode library from each sample and is ready for the next preparation steps (PCR index, second PCR clean-up, library quantification, normalization, pooling, and MiSeq sequencing) or to be sent to the laboratory where these steps will be carried out.

3.8 Index PCR

In this step a unique identifier, or index sequence, is added to each sample via a short 8-cycle PCR step (*see* **Note 41**). Once libraries are indexed, multiple libraries can be pooled and sequenced together. In brief, we prepared dual-indexed libraries with two 8-base pair Nextera XT Index Primers to generate unique tagged libraries as described in the Illumina protocol [16].

3.9 PCR Clean-up 2

The procedure is very similar to Subheading 3.7 except for a larger starting PCR volume of 50 μL, the volume of the reagents to be added and the final volume of the purified libraries (*see* Table 8). To clean up a full 96-well plate with 50 μL PCRs, approximately 5.5 mL of beads will be required (*see* **Note 11**).

Table 8
PCR clean-up 2 consumables and quantities

Name	Quantities	Storage
AMPure XP beads	56 μL per sample	Aliquots of 2.5 and 3.0 ml prepared and stored in the fridge at 2° to 8 °C
80% ethanol	400 μL per sample	200 mL freshly prepared
10 mM Tris pH 8.5	27.5 μL per sample	Aliquots of 50 mL prepared and stored between −15° and −25 °C. Aliquots can be stored up to 6 months
PCR reservoir for multichannel 25 mL	3	
PCR seal sheets	3	
96-well plate	1 plate	

After the second PCR clean-up, the PCR plate contains purified and indexed sequencing libraries, and it is ready for quantification, normalization, pooling, and MiSeq sequencing or to be sent to the laboratory where these steps will be carried out (*see* **Note 42**). Store plate at −20 °C for up to 1 week (safe stop point).

3.10 Library Quantification: PicoGreen Assay

For library quantification, it is recommended to use a fluorometric quantification method based on dsDNA binding dyes (PicoGreen or Qubit assay). To quantify concentrations of the purified indexed sequencing libraries, analyze 6 μL using a Quant-iT PicoGreen assay that uses a fluorescent nucleic acid stain. Samples are read with a standard fluorescent microplate reader and the results returned are in the ng/μL range (*see* **Note 43**).

3.11 Determine the Library Size: Bioanalyzer

To check the amplicon sizes, 2 μL of 11 randomly selected samples is run on a Bioanalyzer DNA 2100 (*see* **Notes 42** and **43**). The results are then used to estimate the average library size which, in this ITS1 metabarcoding assay, is approximately 400 bp.

3.12 Manual Normalization of Libraries

The efficiency of barcode amplification varies from one sample to the next so, in this step, the concentration of each sample's amplicon library is standardized to a single nM value by factoring the PicoGreen results from Subheading 3.10 (returned in ng/μL) with the average library size calculated from Subheading 3.11 multiplied by 660 g/mol (the average molecular mass of 1 base pair) as described in the Illumina protocol [16]. A new 96-well plate is generated in which individually calculated volumes of 10 mM Tris pH 8.5 and equal volumes of each amplicon library from Subheading 3.9 are added (*see* **Note 44**).

4 Notes

1. Plan the work with care, considering that the detection methods are very sensitive, and it is vital to limit cross-contamination risks at all stages in the process. Special attention should be given to sample labels, as these Sample IDs will follow the samples from the field to the bioinformatic analyses. Standardized metadata is key to data organization and downstream analysis.
2. Longmire preservation buffer will form a sediment at lower temperatures. If a white crystallized sediment is observed at the bottom of the bottle, place in a water bath or heating block at 65 °C for a few minutes until the buffer is clear prior to dispensing into 15 mL vials.
3. Dissolve one tablet of Palintest Instachlor directly in the knapsack containing 5 L of tap water to give a final concentration of 30 mg/L. Agitate to dissolve and pump some water through the hose to ensure that the solution is dispersed thoroughly. Removable parts of the knapsack and filter holder should be placed in a bucket containing the same volume and concentration of chlorine. Leave to stand for 30 min (contact time). Flush out the chlorine solution through the knapsack to empty it. Rinse all equipment with clean water prior to use. A 10% solution of domestic bleach could also be used for disinfection of the knapsack and other tools used in the field with the same contact time as described above. However, during agitation of the solution in the knapsack, foam is produced, and larger volumes of water are needed to flush out the equipment before further use.
4. The knapsack sprayer is modified by replacing the trigger and lance with a poly-propylene filter holder into which the filters are fitted (Fig. 1b). Identify the knapsack by number (ID) if more than one in use.
5. We label samples as follows: Acronym of the sampling site (XX) followed by yy/mm/dd and sample rep ID (A to C) corresponding to the three replicate samples from the same sampling point.
6. Cut rectangular pieces of muslin (approximately 20 × 45 cm) and store inside a plastic bag. Fold 1 piece twice and place over the strainer in the lid to filter out coarse debris.
7. Prepare 400 mL of fresh 10% bleach for laboratory disinfection prior to starting the work.
8. Kapa is a proofreading *Taq* polymerase used for all PCRs. It minimizes PCR errors and is suggested in the Illumina protocol.

9. Prepare 1 L of 10 mM Tris–HCl pH 8.5 and autoclave. Once cooled make 50 mL aliquots and label vials with concentration, pH, date, and store at −20 °C for up to 6 months.
10. Prepare 200 mL of 80% ethanol fresh each time. Ethanol is hygroscopic, and over time, working stocks absorb water from the atmosphere, lowering their concentration.
11. When a new bottle of AMpure XP beads is opened, vigorously vortex the whole container to resuspend the beads. Prepare working aliquots of 2.5 mL and 3.0 mL by dispensing into 7 mL bijou containers labelled with date and volume and store in the fridge at 5–8 °C. To clean up a full 96-well plate with 25 μL PCR reactions, approximately 2.0 mL of beads will be needed.
12. A dedicated set of pipettes, glassware (bottles, measuring cylinders, etc.), and clean lab coats should be used for preparing metabarcoding libraries.
13. Pre-filter muslin cloth is used if water has debris that would quickly block the filter.
14. Do not use excessive pressure as damage to the filter membrane will cause loss of material. In our experience, a single filter is sufficient for clean water and three for water with sediment. To prevent cross-contamination, gloves should be worn when handling filters and changed between replicate samples at the same site and between sites.
15. It is crucial that filters are placed in the vial with buffer immediately after filtration and fully submerged, as eDNA can degrade rapidly. We use clean pipette tips to submerge the filters. Vials with filters should be stored in the cooler with frozen ice packs if there will be a long delay between sampling and processing in the laboratory or on particularly warm days.
16. When tap/clean water is not available, the disinfection procedure may be carried out after filtration using water from the sampling point with a chlorine tablet added. Prior to starting the sampling from the new point, the chlorinated water is discarded, and the system flushed with water from the new sampling point. Having more than one knapsack sprayer is advisable.
17. Daily laboratory disinfection procedures on working areas, tools, and equipment are vital to minimize the risk of DNA contamination. Wipe surfaces of benches, centrifuges, and pipettes with 10% bleach. Racks should be submerged in 10% bleach for 15 min, washed, and dried.
18. Tubes with samples in preservation buffer can be left in the fridge at 5 °C overnight if filters are fully submerged.

19. For eDNA elution, 50 μL of the buffer AE has been used instead of 100 μL, and we do not repeat the elution step recommended by the manufacturer.
20. A pre-screen PCR test with a reduced 12.5 μL volume is recommended to identify the *Phytophthora*-positive samples and to minimize costs (Table 3).
21. In our experience, forward planning is critical before starting to run the PCRs with positive samples and subsequent PCR product clean-up with beads. These are intense and laborious procedures, and all materials and solutions required should be prepared in advance and placed within easy reach. Attention should be given to labelling and storing PCR plates correctly.
22. For preparation of the ITS1 metabarcoding libraries from eDNA samples, we follow the Illumina protocol for 16S Metagenomic Sequencing Library Preparation with minor changes, and the adapters tagged to the *Phytophthora* primers ensure compatibility with the Illumina index and Illumina MiSeq System for sequencing.
23. A small volume master mix reaction for eight samples with ITS6F and 5.8S-1R primers was prepared for the synthetic DNA controls. The dilution mix (template DNA) is added to the PCR master mix prior to dispensing the required volume in the designated wells. The synthetic DNA control we use does not have a lead sequence matching the 18Ph2F primers; hence, the first round PCR is carried out with ITS6F and 5.8S-1R primers. An identical number of cycles is used for synthetic DNA control and "real" samples, and the presence of any synthetic DNA reads in "real" sample libraries is used to track any possible contamination during the whole procedure.
24. Run the remaining PCR product that was stored in the fridge on a 2% agarose gel after the PCR clean-up step. This step is considered useful because it allows comparison of the sample amplification intensities and removal or replacement of samples that did not amplify.
25. PCR conditions were optimized based on those of Scibetta [6].
26. Before preparing metabarcoding libraries with "real" positive samples, it is prudent to run an amplicon PCR clean-up practice using PCR product from excess or dummy samples. In this way staff are forewarned of difficulties that can be experienced. Another option is to break the task down to fewer columns at a time (e.g., three columns). Just ensure the Amplicon PCR clean-up of the whole plate is completed on the same day. The amplicon PCR plate to be cleaned must be kept on ice throughout the whole step. Do not freeze and thaw the Amplicon PCR plate. Plan the work in advance and proceed calmly.

27. Turn the bijou container with the AMpure beads working aliquot over several times to dislodge the beads sedimented at the bottom of the container. Vortex for 30 s or until a homogeneous color is observed throughout.
28. Pour the 2.5 mL aliquot of beads into the reservoir container. To ensure that beads are evenly dispersed in the reservoir, pipette up and down twice each time before taking 20 μL to dispense into each well. If bead droplets form on the outside of the pipette tips, remove the droplets prior to adding to the wells.
29. A homogeneous rusty color will be observed in all wells across the plate. If any colour gradients are observed, then mix again. By gently mixing the beads, cross-contamination splashes between wells, formation of "beads" droplets near the top of wells and formation of air bubbles at the bottom of the wells can be avoided/minimized.
30. The amplicon PCR plate should have a visible pellet of rusty coloured beads attached to the wall of each well on the magnetic side of the plate. Do not remove the sample plate from the magnetic stand.
31. If beads are aspirated by accident, dispense everything back very slowly into the wells of the same column, avoiding splashes and discard tips. Leave the beads from this column to re-form the pellet on the magnetic side of the plate. Meanwhile, continue with the removal of the supernatant from the subsequent columns. Once the supernatant has been removed from all other columns, return to the column left behind to remove the supernatant.
32. Before starting the wash steps, we recommend having four full boxes of 200 μL filtered pipette tips ready for all eventualities such as tips falling off from the multichannel pipette. During these steps, it is important to prepare well and work reasonably quickly.
33. Pour 20 mL of freshly prepared 80% ethanol into a new reservoir container. Refill the ethanol reservoir as necessary. The same ethanol reservoir is used for the two wash steps. Pipette ethanol gently to minimize contamination caused by splashes between wells and the formation of "ethanol" droplets near the top of the well. The presence of large droplets at the top of the wells can contribute to ethanol carryover during the elution of the pellets. Beware of the low viscosity of ethanol, which may result in drips from pipette tips. Minimize the movement of full tips over the whole 96-well plate by placing the reservoir above or below the plate rather than to the right or left. Discard the ethanol reservoir after the second wash.

34. Great caution must be taken during this step to avoid aspirating beads when removing the ethanol (*see* **Note 31**).
35. Keeping the amplicon PCR plate on the magnetic stand, remove the supernatant and discard. If necessary, after removing the supernatant, the plate can be briefly removed from the magnetic stand to check if all ethanol has been removed from the bottom of all wells, principally those from the middle wells of columns 5–8, which are more difficult to see clearly from the side. Return the plate immediately to the magnetic stand.
36. Leave the pellet to air dry, and make sure that there are no ethanol droplets on the wall at the top of the wells or at the bottom of the tube. Do not leave pellets to fully dry as this can reduce DNA yield. As a guide, if you see shiny pellets, they are still wet, but if small cracks are observed in the pellets, they are drying out too much, and you should proceed to the next step as rapidly as possible.
37. The same tips can be used to dispense the elution buffer throughout the whole plate, providing great care is taken to avoid splashes and touching the bead pellets.
38. If pellets are firmly adhered to the well wall, do not rub with pipette tips but seal the amplicon PCR plate and vortex vigorously for 30 s to dislodge pellets. Centrifuge the amplicon plate and gently pipette the entire volume up and down ten times (*see* **Note 28**). Discard tips between columns. Check for the presence of air bubbles. If present, remove air bubbles by gently mixing the whole content of individual wells with a single-channel pipete set at 45 μL.
39. If you are concerned about carrying over beads, place the plate with the newly transferred eluate back on the magnetic stand for 2 min. Then, check for the presence of rusty-colored bead pellets on the magnetic side of the stand. If pellets are present, carefully transfer the whole eluate to a new plate.
40. For those preparing libraries with eDNA samples for the first time, it is suggested to run PCR products from before and after the first PCR clean-up on a 2% agarose gel. This is very helpful because (1) it allows visualization of the same sample's amplification intensity in both steps; (2) samples that did not amplify or with no clearly visible amplification after the PCR clean-up can be removed and replaced. Some samples may have limited amplification after the clean-up step, but this may improve after the eight Index PCR cycles (Subheading 3.8); Illumina suggest running some of the samples on the Bioanalyzer, to check sizes. However, as we know the approximate amplicon size, we prefer to run the samples on a gel and only send selected samples to the Bioanalyzer after the second PCR clean-up.

41. Prior to setting up the reactions, prepare a template spreadsheet with all sample names and add the name of the Nextera Indexes. As the final volume for each Index PCR reaction is 50 μL, then 35 μL of master mix is needed per reaction, requiring a 5 mL Eppendorf to accommodate the 3.5 mL of master mix needed for 100 reactions. The master mix is prepared with PCR grade water and KAPA HiFi HotStart ReadyMix and distributed into each well before adding the PCR amplicon and lastly the index primers.
42. Run the 3 μL from the second PCR clean-up on a 2% agarose gel and compare gel results with those from the first PCR-clean up. The gel also can be used to visualize amplicon size variation, which helps select the 11 samples for library size checking on the Bioanalyzer.
43. Before aliquoting DNA samples for PicoGreen and Bioanalyzer assays, we recommend contacting the genomic laboratory where the samples will be run to check the guidelines for sample submission (volume and concentrations). Samples are submitted in a PCR plate with the template ID. We use the PicoGreen assay as it is suited for the quantification of larger numbers of up to 192 libraries for each plate. For fewer libraries, the Qubit quantification assay can be used. For Qubit assays, follow the protocol recommended by the Qubit fluorimeter manufacturer.
44. Before starting to normalize libraries, plan dilutions carefully. Also, we recommend contacting the laboratory that will run the libraries on the MiSeq for their guidance on pooled library final dilution. The Illumina protocol recommends diluting the final library concentration to 4 nM, but this may vary.

References

1. Ruppert KM, Kline RJ, Rahman MS (2019) Past, present, and future perspectives of environmental DNA (EDNA) metabarcoding: a systematic review in methods, monitoring, and applications of global EDNA. Glob Ecol Conserv. https://doi.org/10.1016/j.gecco.2019.e00547
2. Renshaw MA et al (2015) The room temperature preservation of filtered environmental DNA samples and assimilation into a phenol–chloroform–isoamyl alcohol DNA extraction. Mol Ecol Resour 15:168–176. https://doi.org/10.1111/1755-0998.12281
3. Minamoto T et al (2012) Surveillance of fish species composition using environmental DNA. Limnology 13:193–197. https://doi.org/10.1007/s10201-011-0362-4
4. Spens J et al (2017) Comparison of capture and storage methods for aqueous macrobial eDNA using an optimized extraction protocol: advantage of enclosed filter. Methods Ecol Evol 8:635–645. https://doi.org/10.1111/2041-210X.12683
5. Matsuoka S, Sugiyama Y, Nagano M, Doi H (2022) Influence of DNA extraction kits on freshwater fungal DNA metabarcoding. PeerJ 10:e13477. https://doi.org/10.7717/peerj.13477
6. Scibetta S et al (2012) A molecular method to assess *Phytophthora* diversity in environmental samples. J Microbiol Methods 88:356–368. https://doi.org/10.1016/j.mimet.2011.12.012
7. Prigigallo MI et al (2015) Molecular analysis of *Phytophthora* diversity in nursery-grown

ornamental and fruit plants. Plant Pathol 64: 1308–1319. https://doi.org/10.1111/ppa.12362

8. Catala S et al (2015) The use of genus-specific amplicon pyrosequencing to assess *Phytophthora* species diversity using eDNA from soil and water in Northern Spain. PLoS One 10(3). https://doi.org/10.1371/journal.pone.0119311
9. Catala S et al (2017) Metabarcoding and development of new real-time specific assays reveal *Phytophthora* species diversity in holm oak forests in eastern Spain. Plant Pathol 66:115–123. https://doi.org/10.1111/ppa.12541
10. Khaliq I et al (2018) eDNA from roots: a robust tool for determining *Phytophthora* communities in natural ecosystems. FEMS Microbiol Ecol 94:fiy 048. https://doi.org/10.1093/femsec/fiy048
11. Green S et al (2021) PHYTO-THREATS: addressing threats to UK forests and woodlands from *Phytophthora*; identifying risks of spread in trade and methods for mitigation. Forests 12:1617. https://doi.org/10.3390/f12121617
12. La Spada F et al (2022) DNA Metabarcoding and isolation by baiting complement each other in revealing *Phytophthora* diversity in anthropized and natural ecosystems. J Fungi 8:330. https://doi.org/10.3390/jof8040330
13. Muha PT (2019) An optimised eDNA protocol for detecting fish in lentic and lotic freshwaters using a small water volume. PLoS One 14(7). https://doi.org/10.1371/journal.pone.0219218
14. Qiagen (2020) DNeasy Blood & Tissue Handbook. https://www.qiagen.com/us/resources/resourcedetail?id=68f29296-5a9f-40fa-8b3d-1c148d0b3030&lang=en. Accessed 28 June 2023
15. Roche (2022) KAPA HiFi HotStart ReadyMix PCR Kit Technical Data Sheet – KR0370 – v14.22. https://rochesequencingstore.com/wp-content/uploads/2023/02/KAPA-HiFi-HotStart-ReadyMix-PCR-Kit-Technical-Data-Sheet_v14-22.pdf. Accessed 28 June 2023
16. Illumina (2010) 16S Metagenomic Sequencing Library Preparation Preparing 16S Ribosomal RNA Gene Amplicons for the Illumina MiSeq System. https://support.illumina.com/documents/documentation/chemistry_documentation/16s/16s-metagenomic-library-prep-guide-15044223-b.pdf. Accessed 28 June 2023
17. Beckman Coulter (2016) Agencourt AMpure XP PCR purification Instructions for Use B37419AB. https://www.beckmancoulter.com/wsrportal/techdocs?docname=b37419. Accessed 28 June 2023

Chapter 10

A Method for the Separation of *Phytophthora* Oosporesfrom Soil for DNA-Based Detection

Jade T. T. Palmer and Monica L. Gerth

Abstract

Here, we present a protocol for the isolation and detection of *Phytophthora* oospores directly from soil samples. Our method incorporates a novel technique for isolating *Phytophthora* oospores using filter pouches and an improved DNA extraction procedure specifically designed for oospores. While we have primarily developed this protocol for detecting *P. agathidicida* oospores using end-point PCR, we believe these methods can be readily adapted for other *Phytophthora* species. Furthermore, the DNA extracted using this protocol is suitable as input for other DNA-based detection methods.

Key words Oospores, *Phytophthora agathidicida*, Soil, Filtration, DNA, PCR, Detection

1 Introduction

Phytophthora species are destructive plant pathogens that pose significant threats to crops, forests, and natural ecosystems worldwide. Early and accurate detection of *Phytophthora* in soil is crucial for implementing effective disease management strategies and preventing widespread outbreaks. Traditionally, detection methods for *Phytophthora* in soil have relied on culture-based techniques, such as baiting and plating on selective media [1–3]. While these methods have been widely used, they are often time-consuming and labor-intensive, and slow-growing or fastidious species can be easily overlooked or outcompeted by faster-growing organisms during the culturing process [3–7].

In recent years, DNA-based detection methods such as Polymerase Chain Reaction (PCR), quantitative PCR (qPCR), and Loop-mediated Isothermal Amplification (LAMP) have gained prominence due to their increased speed, sensitivity, and specificity. The use of DNA-based methods for the detection of *Phytophthora* from bait plant material [8–12] reduces the time and associated labor costs. However, due to the baiting step, many of the same

Monica L. Gerth and Rosie E. Bradshaw (eds.), *Phytophthora: Methods and Protocols*, Methods in Molecular Biology, vol. 2892, https://doi.org/10.1007/978-1-0716-4330-3_10,

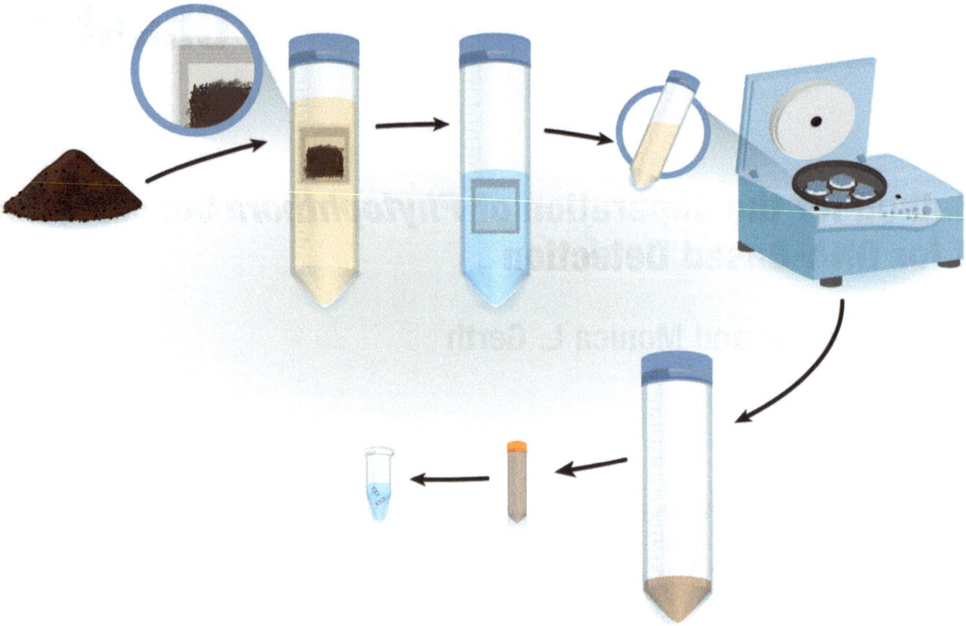

Fig. 1 Overview of the protocol for extracting DNA from oospores from soil samples and preparing them for molecular analysis. The process begins by placing 5–10 g of soil in a double-filter pouch. When soaked in water, oospores separate from small and large fractions of the bulk soil, remaining in the outer pouch. This outer pouch is then rinsed and centrifuged, resulting in a pellet containing oospores, any other similarly sized particles and residual soil. This pellet undergoes an optimised DNA extraction process, followed by downstream DNA detection methods such as PCR

limitations remain. Researchers have also explored DNA-based methods for the direct detection of *Phytophthora* in soil samples, but these approaches face several challenges. These challenges include issues associated with the volume of soil needed for detection [13, 14], ineffective oospore lysis [15] and/or the co-extraction of soil compounds that inhibit downstream PCR reactions [16, 17]. Given these limitations, there is a pressing need for improved techniques that can reliably and efficiently detect *Phytophthora* in soil.

Here, we present a method for detecting *P. agathidicida* oospores in soil using PCR. *P. agathidicida* is a soil-borne pathogen of kauri trees in Aotearoa, New Zealand. It is a homothallic species whose ability to form long-lived oospores is a key part of its life cycle [18].

An overview of our method is shown in Fig. 1. The first step involves separating oospores from bulk soil and other organisms based on size using a double-membrane filter pouch, inspired by protocols for isolating intracellular DNA from marine sediment [19]. *P. agathidicida* oospores, like those of other *Phytophthora* species, are approximately 35 μm in diameter [20]. This size is larger than most bacterial cells and free environmental DNA [21] but smaller than coarse soil particles and organic matter. The filtration step has the added benefit of removing many of the inhibitory compounds present in soil. The oospores and residual soil are then pelleted, and the DNA is extracted using a bead-beating lysis method. Finally, the extracted DNA is used in an end-point PCR with positive, negative, and inhibition controls.

We have developed this protocol for detecting *P. agathidicida* oospores using end-point PCR with species-specific primers, which were designed based on the organism's genome sequence [22]. While our focus has been on *P. agathidicida*, we believe these methods can be readily adapted for other *Phytophthora* species. Furthermore, the DNA extracted using this protocol is compatible with various DNA-based detection methods, including qPCR. When combined with qPCR, our approach can provide valuable information about the concentration of oospores in soil, offering insights that traditional baiting techniques alone cannot provide.

2 Materials

2.1 Equipment

1. Heat sealer (e.g., Mercier 200 mm Hand Impulse sealer).
2. Centrifuge that can accommodate 50 mL tubes and spin up to 3000× *g*.
3. Bento Lab (a portable workstation that includes a PCR machine, a microcentrifuge, a gel electrophoresis system and a blue LED transilluminator) or equivalent equipment (*see* **Note 1**).
4. Vortex mixer with adaptor for 2 mL tubes (e.g., Vortex-Genie 2 with Qiagen Vortex Adapter catalogue #13000-V1-*24*).

2.2 Sample Preparation

1. 5–10 g of soil sample (*see* **Note 2**).
2. Filtered pipette tips.
3. 10% bleach solution (prepare fresh).
4. 90 and 25 μm nylon mesh filter pouches (*see* **Notes 3** and **4**).
5. 50 mL conical centrifuge tubes.
6. Tweezers.
7. 2 mL SK38 soil grinding tubes (Bertin) (*see* **Note 5).**
8. DNeasy PowerSoil Pro Kit (Qiagen).
9. DNeasy PowerClean Pro Cleanup Kit (Qiagen). Optional (*see* **Note 6**).

2.3 PCR and Agarose Gel Electrophoresis

1. 0.2 mL PCR tubes.
2. *Taq* Polymerase PCR master mix (e.g., *Taq* 5× Master Mix (New England BioLabs)).
3. Forward and reverse primers specific for the target organism. For *P. agathidicida*, we use Pa-LTR-for: 5′ ACGCGCTCTG TTTCTTTAGC 3′ and Pa-LTR-rev: 5′ GCGGGCTTCCATT CAATTCA 3′ (*see* **Note 7**).
4. PCR Enhancer Cocktail P (PEC-P; DNA Polymerase Technology). Optional, but recommended (*see* **Note 8**).
5. Positive control DNA: Target organism genomic DNA at a concentration at or higher than the limit of detection. For PCRs using the Pa-LTR primer pair, we use purified genomic DNA at 100 fg/μL for the positive control.
6. Agarose.
7. 50× Tris-acetate EDTA (TAE) buffer: 50 mM EDTA, 2 M Tris, 1 M acetic acid. 1 L of 50×: 242 g Tris-base, 57.1 mL 100% glacial/acetic acid, 100 mL 0.5 M EDTA (pH 8.0), water to 1 L.
8. 1× running buffer for agarose DNA electrophoresis: Dilute 20 mL of 50× TAE into 1 L water.
9. DNA ladder for agarose gel electrophoresis (e.g., Low Molecular Weight DNA ladder).
10. DNA loading dye for PCR samples (if not incorporated into polymerase mix, e.g., Gel Loading Dye, Purple (6×) (New England Biolabs)).
11. GelGreen Nucleic Acid Gel Stain or other appropriate DNA gel stain.
12. DNA surface decontaminant (e.g., bleach or DNA AWAY (Thermo Fisher Scientific)).

3 Methods

3.1 Sample Preparation

1. Clean the work area, tools, and equipment with 10% bleach before and after handling samples.
2. Transfer ~5–10 g of soil into a 90 μm filter pouch until one-half to two-thirds full. Heat seal the open edge of the 90 μm pouch.
3. Place the sealed 90 μm filter pouch inside a 25 μm filter pouch. Heat seal the final edge of the 25 μm outer pouch and place it into a 50 mL centrifuge tube (*see* **Note 9**).

3.2 Concentration of Oospores from the Bulk Soil via Double-Filtration

1. Add sterile water to the sample tube up to 50 mL.
2. Vortex the tube vigorously, then leave it to soak for 30 min.
3. Tip out the water, leaving the double-filter pouch inside the centrifuge tube.
4. Add sterile water to the 50 mL mark. Vortex and shake well, then leave to soak for an additional 15 min.
5. Using clean tweezers, remove the double-filter pouch from the tube. Then, use clean scissors to cut the top edge of the 25 μm (outer) pouch (*see* **Note 10**). Use tweezers to remove the intact 90 μm (inner) filter pouch and discard.
6. Carefully tip the 25 μm filter pouch containing the oospores upside down into a clean 50 mL centrifuge tube, including any liquid remaining in this 25 μm pouch at this stage (*see* **Note 11**).
7. Add sterile water to the tube containing the 25 μm filter pouch and residual liquid to the 25 mL mark on the tube. Vortex and shake well until the pouch is visibly clean (e.g., 15–30 s). Remove the pouch with clean tweezers and discard.
8. Centrifuge the tube of oospore-soil slurry at 3000× *g* for 2 min. Tip or pipette off the liquid, leaving just the pellet.

3.3 Extraction of DNA

The procedures for DNA extraction from soil generally follow the protocols described in the Qiagen DNeasy PowerSoil Pro Kit, with the modifications described below.

1. Resuspend the pellet in 800 μL of CD1 buffer from the DNeasy PowerSoil Pro DNA extraction kit. Transfer the liquid to a SK38 bead beating tube (*see* **Note 5**).
2. Homogenize the sample for 10 min at high speed using a benchtop vortex equipped with a horizontal adaptor.
3. Follow the remaining steps of the PowerSoil Pro Kit as per the manufacturer's instructions. Elute the DNA into 100 μL of C6 solution (*see* **Note 12**).
4. Optional: Further purify the DNA sample using the Power-Clean Pro Cleanup kit (*see* **Note 6**). Elute the DNA into 100 μL of the supplied EB buffer solution or preferred elution buffer.

3.4 PCR

1. Prepare the PCR reactions in thin-walled 0.2 mL tubes. We recommend using two different volumes of input DNA (*see* **Note 13**) and always including positive, negative, and inhibition controls (*see* **Note 14**). If analyzing multiple samples, it can be convenient to set up a master mix with all reagents and divide it among 0.2 mL PCR tubes before adding the sample or control DNA to each tube. See Table 1.

Table 1
Components of the control and sample PCRs

Reagent (per 20 μL reaction)	Positive control (+) (μL)	Negative control (−) (μL)	3 μL soil DNA input (A) (μL)	1 μL soil DNA input (B) (μL)	Inhibition control (IC) (μL)
Taq 5X Mastermix	4	4	4	4	4
Forward primer (10 μM)	1	1	1	1	1
Reverse primer (10 μM)	1	1	1	1	1
PEC-P	10	10	10	10	10
Extracted soil DNA	0	0	3	1	1
Genomic DNA[a]	1	0	0	0	1
Water (to 20 μL final volume)	3	4	1.0	3	2

[a]for positive and inhibition control reactions only

Table 2
Recommended PCR cycling conditions

Denature	95 °C 95 °C	30 s 22 s	×40 cycles
Anneal	55 °C	30 s	
Extend	68 °C 68 °C 4 °C	7 s 5 min Hold	

2. We recommend the thermocycling conditions for the primer pair Pa-LTR-for and -rev seen in Table 2. These conditions may need to be optimised when used with other primer pairs.

3. Run 10 μL of each PCR product with loading dye on a pre-stained agarose gel alongside an appropriate DNA ladder. Our *P. agathidicida*-specific primer pair produces a PCR product of ~161 bp (*see* **Note 15**). For PCR products of this size, we recommend a 2% agarose gel and 100 bp or Low Molecular Weight DNA ladder.

3.5 Interpretation of PCR Results

1. The expected product size will vary depending on the primer set used. The product size for the PA-LTR primer pair is ~161 bp (*see* **Note 15**).
2. Positive control: A single band of the correct size indicates that the PCR reaction worked as intended, even if none of the soil samples produced PCR products. No band indicates the PCR failed, likely due to missing/inactive reagents.
3. Negative control: A band in the negative control (excluding residual primers) indicates contamination in your PCR mix or workflow. Primary sources of contamination include reagents, equipment, or the workspace (*see* **Note 16**).
4. Inhibition control: The presence of a single band of the expected size indicates that the soil DNA sample is free from significant PCR inhibitors. The absence of a band or reduced amplification intensity compared to the positive control suggests the presence of inhibitory compounds in the soil DNA sample. In such cases, negative results in the corresponding sample may be due to inhibition rather than the absence of target DNA. When inhibition is detected, additional purification steps may be necessary to obtain reliable results. These steps could include increasing the ratio of PEC-P or diluting the DNA sample to reduce the concentration of inhibitors. It's important to note that while dilution may reduce inhibition, it may also decrease the concentration of target DNA, potentially affecting assay sensitivity.
5. Interpreting sample results. Assuming all control reactions performed as expected, interpret the sample results as follows. Positive result: The presence of a single, clear band of the correct size in either or both input DNA volumes (1 and 3 μL) indicates that the target DNA is present. This confirms a positive result for the target organism. Negative Result: The absence of a band in both input DNA volumes suggests that the sample is negative for the target DNA. Inconclusive Result: If the controls fail, or the results are inconsistent between the two input volumes, or if unexpected bands appear, the assay may need to be repeated (*see* **Note 17**).

4 Notes

1. A Bento Lab provides a portable and cost-effective experimental setup. However, standard laboratory equivalents of the individual components are equally suitable for conducting the procedures described herein.

2. The samples should consist mainly of soil (i.e., avoid rocks and large pieces of plant material), and the soil should be well homogenized.
3. These filter pore sizes have been selected for *P. agathidicida* oospores, which are approximately 35 μm in diameter. For different *Phytophthora* species and/or oospores of other sizes, the pore sizes may need to be optimised. Optimal pore size can be assessed using microscopy.
4. To prepare the pouches, cut the filter membranes into rectangles of the following dimensions: the 90 μm membrane should measure approximately 12 × 6 cm, and the 25 μm membrane should measure approximately 15 × 8 cm. Fold each rectangle in half to form a square pouch. Heat seal any two edges of each membrane, leaving one side open, with 0.5–1.0 cm margins.
5. The SK38 beads are required to lyse *P. agathidicida* oospores effectively; the beads supplied in the PowerSoil Pro extraction kit are not effective at oospore lysis. It's important to note that using SK38 beads results in highly sheared DNA. While this has not been problematic for our downstream PCR-based analyses, if less fragmented DNA is required, a shorter and gentler bead-beating step can be explored. If the lysis step is modified or used with spores from a different *Phytophthora* species, complete lysis should be visually confirmed by microscopy before proceeding with further steps.
6. Not all soil samples require the additional PowerClean Pro cleanup step. We only use this extra purification step when PCR inhibition is observed in the inhibition control (IC) spiked sample reaction. The extra cleanup step could be integrated into the standard workflow; however, this would increase the time and cost per sample.
7. Primers should be specific to the target organism(s) of interest. Other considerations in primer design are the desired amplicon length and target copy number. The Pa-LTR primer pair targets a repetitive sequence present hundreds of times in the *P. agathidicida* 3770 genome [22], but not that of other sequenced *Phytophthora* species.
8. We find that PEC-P is required to reduce PCR inhibition in our soil samples. However, for soils with low inhibitor amounts, PEC-P could be replaced by water in the PCR reactions.
9. The filter pouches should slide into the tube with some slight pressure. If met with too much resistance, the pouch is overfull and subsequent soaking will be difficult/insufficient.
10. Always cut the top edge of the double filter pouch. There will be liquid remaining in the 25 μm pouch which contains oospores. Therefore, all the remaining liquid should be transferred

to the new clean tube to be included in the next rinsing and pelleting steps. Cutting the sides or bottom will result in loss of the sample liquid and, therefore, oospore loss.

11. The 25 μm pouch should be transferred to a clean 50 mL tube upside down to ensure proper rinsing in the next step. If the pouch is added right side up, clean water will not adequately rinse the pouch, which may result in the loss of oospores.
12. We recommend using 100 μL to elute the samples. This may seem counterintuitive, as it dilutes the DNA sample; however, it also dilutes any inhibitors and, therefore, often increases the overall yield of the PCR product.
13. We find that, typically, 1 μL of PCR input is sufficient to detect target DNA from a 5–10 g sample. However, in some cases, the amount of target DNA may be too low for detection with this volume. To prevent false negatives that can occur with only 1 μL input, we also recommend using a 3 μL input; this increased volume can improve the likelihood of detecting low-abundance target DNA.
14. The positive control checks for correct PCR setup, ensuring all reagents are present and active. Negative control checks for contamination of the PCR with target DNA. The inhibition controls check for inhibitory components in each soil sample, helping reduce false negatives. By adding a known amount of target DNA to a reaction containing the sample extract, researchers can assess the impact of potential inhibitors like humic acids or tannins.
15. Our primer pair (Pa-LTR) targets a highly repetitive region of the genome. Due to the repetitive nature of this region, slight variations exist among the multiple copies throughout the genome. For instance, our primary target product is 160 or 161 base pairs (bp) in length, with the single base pair difference resulting from insertions in some copies. Additionally, there is occasionally a secondary product of 98 or 103 bp, which arises from a deletion in chromosome 6. This secondary product is typically only observed when amplifying high concentrations of *P. agathidicida* DNA and is not usually seen in the lower concentrations typically found in soil extractions.
16. Preventing PCR contamination is crucial, as free DNA is often present throughout laboratory environments. To minimize this risk, it is essential to implement rigorous cleaning protocols, such as using bleach solutions to clean work surfaces. Keep reagent tubes closed as much as possible and use aliquots of reagents instead of stock solutions. We also recommend physically separating different process stages, e.g., establishing dedicated pre-PCR and post-PCR work areas.

17. For pathogen detection, increasing replication can help avoid false negatives, but the optimal number of replicates depends on factors such as the expected abundance of the target pathogen and/or the complexity of the soil sample.

Acknowledgements

Jade Palmer is supported by a Strategic Doctoral Scholarship co-funded by Tiakina Kauri (Ministry for Primary Industries) and Victoria University of Wellington. Development of this method has also been supported by Te Roroa Iwi via a Vision Mātauranga Capability Grant, funded by the Ministry of Business, Innovation and Employment (New Zealand).

References

1. Burgess TI et al (2021) Towards a best practice methodology for the detection of *Phytophthora* species in soils. Plant Pathol 70(3):604–614
2. Pérez-Sierra A, Jung MH, Jung T (2022) Survey and monitoring of *Phytophthora* species in natural ecosystems: methods for sampling, isolation, purification, storage, and pathogenicity tests. Methods Mol Biol 2536:13–49
3. O'Brien PA, Williams N, Hardy GES (2009) Detecting *Phytophthora*. Crit Rev Microbiol 35(3):169–181
4. Ferguson AJ, Jeffers SN (1999) Detecting multiple species of *Phytophthora* in container mixes from ornamental crop nurseries. Plant Dis 83(12):1129–1136
5. Sarker SR, McComb J, Burgess TI, Hardy GESJ (2020) Antimicrobials in *Phytophthora* isolation media and the growth of *Phytophthora* species. Plant Pathol 69(8):1426–1436
6. Sarker SR, McComb J, Burgess TI, Hardy GESJ (2021) Timing and abundance of sporangia production and zoospore release influences the recovery of different *Phytophthora* species by baiting. Fungal Biol 125(6): 477–484
7. Sarker SR, Burgess TI, Hardy GESJ, McComb J (2023) Closing the gap between the number of *Phytophthora* species isolated through baiting a soil sample and the number revealed through metabarcoding. Mycol Prog 22(6):39
8. Winkworth RC (2020) A LAMP at the end of the tunnel: a rapid, field deployable assay for the kauri dieback pathogen, *Phytophthora agathidicida*. PLoS One 15(1):e0224007
9. Sanfuentes E et al (2022) Detection of *Phytophthora cinnamomi* on declining *Araucaria araucana* forests. For Pathol 52(5):e12765
10. Llorente B et al (2010) A quantitative real-time PCR method for *in planta* monitoring of *Phytophthora infestans* growth. Lett Appl Microbiol 51(6):603–610
11. McDougal RL et al (2021) Molecular detection of *Phytophthora pluvialis,* the causal agent of red needle cast in *Pinus radiata*. J Microbiol Methods 189:106299
12. Liu N et al (2019) A duplex PCR assay for rapid detection of *Phytophthora nicotianae* and *Thielaviopsis basicola*. Plant Pathol J 35(2): 172–177
13. Elliot M et al (2015) An improved method for qPCR detection of three *Phytophthora spp.* in forest and woodland soils in northern Britain. For Pathol 45(6):537–539
14. Than DJ et al (2013) A TaqMan real-time PCR assay for the detection of *Phytophthora* 'taxon Agathis' in soil, pathogen of Kauri in New Zealand. For Pathol 43(4):324–330
15. Lees AK, Sullivan L, Lynott JS, Cullen DW (2012) Development of a quantitative real-time PCR assay for *Phytophthora infestans* and its applicability to leaf, tuber and soil samples. Plant Pathol 61(5):867–876
16. Hargreaves SK, Roberto AA, Hofmockel KS (2013) Reaction- and sample-specific inhibition affect standardization of qPCR assays of soil bacterial communities. Soil Biol Biochem 59:89–97

17. Wang H et al (2017) A re-evaluation of dilution for eliminating PCR inhibition in soil DNA samples. Soil Biol Biochem 106:109–118
18. Bradshaw RE et al (2020) *Phytophthora agathidicida*: research progress, cultural perspectives and knowledge gaps in the control and management of kauri dieback in New Zealand. Plant Pathol 69(1):3–16
19. Medina Caro D et al (2023) An improved method for intracellular DNA (iDNA) recovery from terrestrial environments. MicrobiologyOpen 12(3):e1369
20. Weir BSPE, Anand N, Uchida JY, Pennycook SR, Bellgard SE, Beever RE (2015) A taxonomic revision of *Phytophthora* clade 5 including two new species, *Phytophthora agathidicida* and *P. cocois*. Phytotaxa 205(1):21–38
21. Levin PA, Angert ER (2015) Small but mighty: cell size and bacteria. Cold Spring Harb Perspect Biol 7(7):a019216
22. Cox MP et al (2022) Chromosome-level assembly of the *Phytophthora agathidicida* genome reveals adaptation in effector gene families. Front Microbiol 13:1038444

Chapter 11

Annotation of RxLR Effectors in Oomycete Genomes

Nicholas C. Cauldron, Martha A. Sudermann, Camilo H. Parada-Rojas, and Niklaus J. Grünwald

Abstract

Pathogens have evolved effector proteins to suppress host immunity and facilitate plant infections. RxLR effectors are small, secreted effector proteins with conserved RxLR and dEER amino acid motifs at the N terminus and highly variable C termini and are commonly found in oomycete species. We provide computational approaches to annotate RxLR candidate effector genes in a genome assembly in FASTA format with an available GFF file. Hidden Markov Modeling (HHM) is used in combination with regular expressions to search for RxLR and EER amino acid patterns.

Key words EER, Effector, RxLR, Signal peptide

1 Introduction

Pathogens have evolved effector proteins to suppress host immunity and facilitate plant infections [1]. RxLR effectors are small, secreted effector proteins with conserved RxLR and dEER amino acid motifs at the N terminus and highly variable C termini [2]. The C terminus is thought to be involved in rapid coevolution with plant hosts. The RxLR motif was described after sequences of AVR effector proteins from *Phytophthora sojae*, *Phytophthora infestans*, and *Hyaloperonospora arabidopsidis* were aligned to reveal a conserved motif [3]. The widespread occurrence of effectors and putative effectors with the conserved RxLR motif was then observed following the annotation of *Phytophthora sojae* and *Phytophthora ramorum* genomes [4], and RxLRs have since been documented widely in various oomycete pathogen groups.

Functional annotation of RxLR effectors has often led to mis-annotations, and improved prediction tools have continued to be developed over the last decade. A recent study comparing 31 *Phytophthora* genomes revealed a broad difference in RxLR effector gene numbers, ranging from 46 predicted RxLR effector

Monica L. Gerth and Rosie E. Bradshaw (eds.), *Phytophthora: Methods and Protocols*, Methods in Molecular Biology, vol. 2892, https://doi.org/10.1007/978-1-0716-4330-3_11,

genes in *P. pinifolia* to 1183 predicted in *P. megakarya* [5]. Until recently it was thought that the genus *Pythium* lacked RxLR effectors [6, 7]. However, a revised procedure for calling RxLR effectors that included a more variable dEER motif resulted in the discovery of conserved RxLR effectors in *Pythium* spp. ranging in number from 21 to 78 [8]. This regular expression motif definition is included in the notes for reference and represents the basis for the methods described in this protocol. We provide computational approaches to annotate RxLR candidate genes in a genome assembly in FASTA format with an available GFF file. Hidden Markov Modeling (HHM) is used in combination with regular expressions (regex) to search for the RxLR and -EER amino acid patterns [2, 9].

2 Materials

2.1 Input Genome Assembly

To find cytoplasmic effectors in a genome assembly, the assembly needs to be available in FASTA format, which is output by most assemblers. If planning to annotate all gene models, these annotations should also be available in GFF format. This protocol assumes that the GFF is available, and contains gene, mRNA, CDS, and exon features (*see* **Note 1**). If a GFF is not available, disregard Subheading 3.3 of the protocol and only follow the steps outlined in Subheading 3.2.

2.2 Hardware

All steps can be performed on a personal laptop computer with a minimum of four cores as well as on a high-performance computing infrastructure. Software dependencies can be installed on Linux operating systems. Since some are challenging or impossible to install on Windows or Mac operating systems, we recommend using a Linux-based operating system. If only Windows is accessible, we recommend configuring a Windows Subsystem for Linux.

2.3 Software

The workflow uses many publicly available tools, some with licenses that are obtainable to academic users for no charge. Minor version changes to tools may still be compatible with versions used in this protocol (Table 1). Users who successfully complete the protocol with a newer software version can post an issue on the GitHub repository associated with this protocol (https://github.com/grunwaldlab/oomy_RXLRs). An exception is SignalP (v. 3.0), which is used because of its sensitivity to oomycete effectors [11]. Installing tools and configuring the computing environment is outside the scope of this protocol (Table 1). Read the documentation carefully, including any README files associated with each tool (*see* **Note 2**). Specifically, SignalP and TMHMM are

Table 1
Software dependencies required to complete this protocol. All R packages used are available on CRAN

Command-line tools	
EMBOSS v6.6.0	http://emboss.open-bio.org/
HMMER v3.3.2	http://hmmer.org/documentation.html
AGAT v1.2.0	https://github.com/NBISweden/AGAT
Python 2.7.16	https://www.python.org/download/releases/2.7/
Python 3.11.4	https://www.python.org/downloads/
SignalP 3.0	https://services.healthtech.dtu.dk/services/SignalP-3.0/
TMHMM 2.0c	https://services.healthtech.dtu.dk/services/TMHMM-2.0/
R packages	
Seqinr [10]	*Optparse*
Tidyverse *Tidytext*	*effectR* [9] *Devtools*

challenging to install. To install AGAT successfully, it may be necessary to first create and activate a conda environment for AGAT. Containerized installations of AGAT are also available using Docker or Singularity.

After installation of software, especially the older programs like SignalP and TMHMM, ensure all software dependencies are installed and in your PATH, or specify the paths to the executables when running particular programs.

Provided companion scripts are available on GitHub (https://github.com/grunwaldlab/oomy_RXLRs) and were coded in Bash, Python 3, and R. We recommend recording any modified steps into an RMarkdown file in the RStudio environment, which allows execution of code chunks in R, Bash and Python languages to facilitate reproducibility.

3 Methods

3.1 Set Up Computing Environment

1. Download the GitHub repository (https://github.com/grunwaldlab/oomy_RXLRs) with the necessary scripts and test data.
2. Create a working directory in the desired location on the local computer or computing cluster.
3. Make a subdirectory called "data" within the chosen directory to store assembly and annotation files. Also, copy the scripts directory from the GitHub repository into a chosen directory.

4. Optional: Paste code chunks into an R or Rmarkdown document, or within a text editor. A Rmarkdown template is provided on the GitHub repository.
5. Install software dependencies (*see* Table 1).
6. Occasionally, installing multiple versions of Python into one environment can cause conflicts. Ensure the correct versions of each are found in the path.

```
python -version
# Python 3.6 or greater
python2.7 -version
# Python 2.7.14 - 2.7.16
```

3.2 Obtain Candidate RxLRs from All ORFs

The following steps are performed with a genome assembly file only, regardless of whether other gene calls are already available.

1. Obtain all possible open reading frames (ORFs) from the genome. The minimum size is set in number of nucleotides, and here was set to 210 (70 amino acids) following Haas et al. [12]. The option is set to define an open reading frame as any sequence between start and stop codons. If a start codon resides near the end of a contig in the genome assembly and reaches the end of the contig before a stop codon, the partial ORF is included in the output. If using online submission instead of installed software, ensure that selected boxes match the parameters described. The following code in bash sets up environmental variables that will continue to be used in other bash code chunks and performs the ORF search.

```
# Bash environment
mkdir output_data
genome="assembly.fasta"
isolate=$(basename -s ".fasta" ${genome})
minsize_aa=70 # amino acids
minsize_nt=210 # minsize_aa*3 = number of nucleo-
tides
orfs_name="${isolate}.orfs-min${minsize_aa}long.
start2stop"
orfs="output_data/${orfs_name}.fasta"
# Plug parameters into EMBOSS GetOrf
# Parameter find=1 sets search to between start and
stop codons.
getorf -sequence "$genome" -outseq "$orfs" -minsize
$minsize_nt -find 1
```

2. RxLR effectors are a component of the *Phytophthora* secretome. To obtain a complete "secretome," test all ORFs for evidence of secretion. Evidence includes the presence of a signal peptide, inferred with SignalP (v. 3.0), and the absence of transmembrane helices (TMs), inferred with TMHMM (*see* **Note 3**). SignalP is slow, but the companion Python script parallelizes it by breaking input sequences into chunks. Modify the invocation of python2.7 to use 2–16 threads, depending on resource availability. If needed, modify the command below that invokes the Python script to use the directory with the downloaded script.

```
# Bash environment
threads=4
python2.7 scripts/rxlr_signalpeptide.py $orfs
$threads secretome output_data/${orfs_name}.
rxlr_signalpep.out
tmhmm -short $orfs > output_data/${orfs_name}.tmhmm.
out
```

3. Scan all ORFs for evidence of homology to RxLR proteins using three known motifs (*see* **Note 4**). Next, perform the HMM search to find WY domains often found in C-termini of RxLR proteins (*see* **Note 5**). Then, parse results into an R session. Below, we demonstrate the HMM search using bash scripting, and then parse those search results into the R session.

```
# Bash environment
hmmsearch --cpu 4 --seed 123 -T 0 --tblout output_-
data/${orfs_name}.rxlr_WYfold_hmm.out data/WY_fold.
hmm $orfs > output_data/${orfs_name}.rxlr_WY-
fold_hmm.log
# Simple list of candidates
egrep -v '^#' output_data/${orfs_name}.rxlr_WY-
fold_hmm.out | awk '{print $1}' > output_data/
${orfs_name}.rxlr_WYfold_hmm.list
# R environment
# load pre-installed libraries
source("scripts/secretion_funcs.R")
library("tidyverse")
library("seqinr")
library("tidytext")
library("devtools")
devtools::install_github(repo = "grunwaldlab/
effectR", build_vignettes = TRUE)
library("effectR")
library("optparse")
```

```
wy_list_regex <- "*.orfs-min70long.start2stop.
rxlr_WYfold_hmm.list"
orfs_wys <- list.files(path = "output_data/",
 pattern = wy_list_regex, full.names = TRUE) %>%
 map_dfr(read_wy_list) %>%
 unite("ID_isolate", ID, isolate, sep=":", remove =
FALSE) %>%
 mutate("method" = "WY_hmm")
```

4. All methods of identifying RxLR gene homologs require an RxLR motif, even if degenerate. Here, multiple patterns for RxLR are scanned, including Win2007 [13] and Whisson2007 [2]. The Whisson2007 motif is scanned for both its original complete "RXLR-EER" as well as just "RXLR." To locate EER motifs, we also search for the Whisson2007 motif without the RxLR.

```
# R environment
# Locate files programmatically in R
# all ORFs:
orf_fs_regex <- ".*orfs-min70long.start2stop.fasta"
orf_fs <- list.files(path = "output_data",
 pattern = orf_fs_regex, full.names = TRUE) %>%
 tibble("orf_fs" = .) %>%
 separate(orf_fs, into = c("isolate", NA),
 sep = "(?=\cr.orfs-min)", remove = FALSE) %>%
 mutate(isolate = str_remove(isolate, ".*data/*"))
# SignalP output files:
sp3_fs_regex <- ".*orfs-min70long.start2stop.
rxlr_signalpep.outsp3_tabular.tmp"
sp3_fs <- list.files(path = "output_data",
 pattern = sp3_fs_regex, full.names = TRUE) %>%
 tibble("sp3_fs" = .) %>%
 separate(sp3_fs, into = c("isolate", NA),
 sep = "(?=\cr.orfs-min)", remove = FALSE) %>%
 mutate(isolate = str_remove(isolate, ".*data/*"))
orf_sp3_fs <- full_join(orf_fs, sp3_fs, by = "iso-
late") %>%
 select(orf_fs, sp3_fs, isolate)
# run motif search
allorfs_re_effectr <- map2(orf_sp3_fs$orf_fs,
orf_sp3_fs$sp3_fs, effectr_eer_sp3)
allorfs_re_effectr_tb <- tibble(bind_rows(allorfs_-
re_effectr)) %>%
 unite("ID_isolate", ID, isolate, sep=":", remove =
FALSE)
```

5. Functions in the companion R script remove candidates based on the HMM score calculated in SignalP (v. 3.0) (*see* **Note 6**). The functions used in **step 4** will report the number of hits from SignalP, but some of those proteins do not escape the cell membrane and are, therefore, not a component of the secretome. Since TMs are often a false positive call when reported in a signal peptide region, proteins are only removed if the mean position of at least one TM follows the most likely cleavage site for a mature protein.

```
# R environment
# Programmatically find files in R
sp3_fs_regex    <-     ".*orfs-min70long.start2stop.
rxlr_signalpep.outsp3_tabular.tmp"
sp3_fs <- list.files(path = "output_data",
 pattern = sp3_fs_regex, full.names = TRUE)
tmm_fs_regex  <-  ".*orfs-min70long.start2stop.tmhmm.
out"
tmm_fs <- list.files(path= "output_data",
 pattern = tmm_fs_regex, full.names = TRUE)
# make dataframe for easier function running
# and validate the signalp and tmhmm runs line up
sp3_tmm_fs <- tibble("sp3_fs" = sp3_fs,
 "tmm_fs " = tmm_fs) %>%
 mutate("isolate_s" = str_extract(sp3_fs, "(?<=/).*
(?=\cr.orfs)")) %>%
 mutate("isolate_t" = str_extract(tmm_fs, "(?<=/).*
(?=\cr.orfs)"))
# parse signalP and tmhmm output files.
# filters based on HMM_Sprob_score of signalp
# and position of helix relative to most likely sig-
nal peptide position
sp3_tmm_pass  <-  pmap(sp3_tmm_fs,  ~  read_prune_tms
(..1, ..2, ..3), .id = 'isolate') %>%
 bind_rows() %>%
  unite("ID_isolate", Protein, isolate, sep= ":",
remove = FALSE)
# Size of complete secretome
sp3_tmm_pass %>%
 filter(HMM_Sprob_score >= 0.9) %>%
 distinct(ID_isolate, .keep_all = TRUE) %>%
 group_by(isolate) %>%
 summarize(n_noTM_yesSP = n())
```

6. Filter down final candidates to only those meeting the specified thresholds. This is performed using data frames already created in our R environment.

```
# R environment
# Start with results of RXLR-EER domain searches
rxlr_orfs <- allorfs_re_effectr_tb %>%
 bind_rows(orfs_wys) %>%
 group_by(ID_isolate, ID, isolate) %>%
 mutate(isolate = str_remove(isolate, "/")) %>%
 summarize(methods_list = paste(method, collapse =
",")) %>%
 filter((grepl("Whis2007", methods_list)) |
  (grepl("Whis_rxlr", methods_list) & (grepl("Whi-
s_eer", methods_list) | grepl("WY_hmm", methods_-
list))) |
 (grepl("Win2007", methods_list) & (grepl("Whis_eer",
methods_list) | grepl("WY_hmm", methods_list)))) %>%
 filter(ID_isolate %in% sp3_tmm_pass$ID_isolate) %>%
 group_by(isolate) %>%
 distinct(ID_isolate)
# Final counts
rxlr_orfs %>%
  summarize(noTM_whis_or_whiswinrxlrEER_whiswinrxlr-
WY_count = n())
out_prefix <- "/orfs_cand_RXLRs_"
# Write final candidates to lists using respective
isolate IDs
allorfs_re_effectr_tb %>%
 ungroup() %>%
 distinct(ID_isolate, .keep_all = TRUE) %>%
 select(ID, isolate) %>%
 group_by(isolate) %>%
  group_walk(~ writeLines(.x$ID, paste0("output_-
data", out_prefix, .y$isolate, ".txt")))
```

7. In the bash environment, prepare a GFF file from the RxLR candidates in **step 6** (*see* **Note 7**) to obtain their positional data. This requires the output from above and the original ORF file. Take note of any candidates that overlap each other and add start and stop codons (*see* **Note 8**).

```
# In a bash environment, change directory names if
needed
#Optional commands if AGAT was installed in a conda
environment
#eval "$(conda shell.bash hook)"
#conda activate agat
outdir="output_data"
```

```
# This must be changed depending on location of user's data
assembly_dir="data"
rxlr_prefix="orfs_cand_RXLRs_"
# Run on all genomes being annotated
for orf_rxlr_list in $(ls -1 $outdir/${rxlr_prefix}*.txt); do
 # Establish output file name conventions
 list_name=$(basename --suffix ".txt" $orf_rxlr_list)
 iso_name=$(echo $list_name | sed "s/$rxlr_prefix//")
 echo $iso_name
 assembly=$(ls ${assembly_dir}/${iso_name}.fasta) || \
   assembly=$(ls ${assembly_dir}/other_refs/${iso_name}.fasta)
 # If needed, replace 70 with minimum protein length chosen.
   orfs="$outdir/${iso_name}.orfs-min70long.start2-stop.fasta"
 # Obtain nationality. First get long ORF names from abbreviated.
 grep -w -f $orf_rxlr_list $orfs | sed 's/>//' > $outdir/${list_name}_longnames.list
 # Next convert ORF tag list to GFF
 Rscript scripts/getorf_seqnames2gff.R -i $outdir/${list_name}_longnames.list
 # Notifies about overlapping genes
 agat_sp_fix_overlaping_genes.pl -f $outdir/${list_name}_longnames.gff \
 -o $outdir/${list_name}_longnames.merge_ovlp.gff
 # If genes need merging, pick ORFs by hand or use output from above
   agat_sp_add_start_and_stop.pl --gff $outdir/${list_name}_longnames.gff \
 --fasta $assembly \
 --output $outdir/${list_name}_longnames.str_stp.gff
done
```

8. If no reference GFF is available, rename the final GFF of candidate ORFs to suit your needs. The filename of the final GFF is listed in **Note 9**. Optionally, read and execute code from **step 1** of Subheading 3.2 to rename candidate features with AGAT.

3.3 Obtain Candidate RxLRs from All Coding Genes

This step can be skipped if gene annotations are not available (*see* **Note 1**). The GFF of candidate ORFs obtained from the final step is supplemented into the existing annotation, all proteins are extracted, and **steps 2–6** from Subheading 3.2 will be repeated.

1. Match filename convention of assembly and annotation files, supplement RxLR ORF candidates into the GFF (*see* **Notes 9** and **10**), and extract all predicted proteins. The companion Python script used to complement the GFF with RxLR candidates can be further configured to include additional outputs that may be informative to understand how the total number of RxLRs predicted changes between ORFs and all proteins.

```
#In a bash environment, change directory names if
needed
#Optional commands if AGAT was installed in a conda
environment
#eval "$(conda shell.bash hook)"
#conda activate agat
genome="data/PR-102_v4.fasta"
isolate=$(basename -s ".fasta" ${genome})
outdir="output_data"
assembly_dir="data"
rxlr_prefix="orfs_cand_RXLRs_"
# Run on all genomes being annotated
for orf_rxlr_list in $(ls -1 $outdir/${rxlr_prefix}*.
txt); do
 # Establish output file name conventions
 list_name=$(basename --suffix ".txt" $orf_rxlr_list)
 iso_name=$(echo $list_name | sed "s/$rxlr_prefix//")
 assembly=$(ls ${assembly_dir}/${iso_name}.fasta)||\
   assembly=$(ls  ${assembly_dir}/other_refs/${iso_-
name}.fasta)
 # Make sure that the GFF file name matches the assem-
bly's name. You can use the mv command to make sure
they match. (example: assembly.fasta and assembly.
gff)
   ref_gff=$(ls  $assembly_dir/${iso_name}.gff)  ||
ref_gff=$(ls  $assembly_dir/other_refs/${iso_name}.
gff)
 echo $ref_gff
 # Re-name gene features
 agat_sp_manage_IDs.pl --tair --type_dependent \
 --prefix "${iso_name}_rxlrORF" \
 --nb 1 --gff $outdir/${list_name}_longnames.str_stp.
gff \
```

```
  --output $outdir/${list_name}_longnames.str_stp.re-
IDs.gff
 ls $ref_gff || \
 printf "GFF not found, configure \$iso_name:$iso_name
and \$ref_gff:$ref_gff\" && exit 2"
 echo $ref_gff
 python3 scripts/complement_gff.py -r $ref_gff \
  -s  $outdir/${list_name}_longnames.str_stp.re-IDs.
gff \
  -o  $outdir/${list_name}_longnames.str.stp.re-IDs.
add_rxlrs.gff \
  -l  $outdir/${list_name}_longnames.str_stp.re-IDs.
overlapping_ref.gff \
 --unsorted
 # Extract all proteins including RXLR ORFs supple-
mented
 agat_sp_extract_sequences.pl \
   --gff  $outdir/${list_name}_longnames.str.stp.re-
IDs.add_rxlrs.gff \
 --fasta $assembly \
 --protein \
  -o  $outdir/${list_name}_longnames.str_stp.re-IDs.
add_rxlrs.fasta
done
```

2. Repeat Subheading 3.2, **steps 2–6**, using all proteins. First, test for evidence of secretion by running SignalP and TMHMM in the bash environment.

```
# Bash environment
# change default directory names if needed
genome="data/assembly.fasta"
isolate=$(basename -s ".fasta" ${genome})
peps_name="orfs_cand_RXLRs_${isolate}_longnames.
str_stp.re-IDs.add_rxlrs"
peps="output_data/${peps_name}.fasta"
threads=10 #Change depending on computing capabil-
ities
python2.7    scripts/rxlr_signalpeptide.py    $peps
$threads secretome
output_data/${peps_name}.rxlr_signalpep.out
tmhmm -short $peps > output_data/${peps_name}.tmhmm.
out
```

3. Search for WY domains and parse results into R.

```
# In a bash environment
# Change directory names if needed
hmmsearch --cpu 4 --seed 123 -T 0 --tblout output_-
data/${peps_name}.rxlr_WYfold_hmm.out data/WY_fold.
hmm $peps > output_data/${peps_name}.rxlr_WY-
fold_hmm.log
#remove slash before $1, when working with R bash
code chunk
egrep -v '^#' output_data/${peps_name}.rxlr_WY-
fold_hmm.out | awk '{print $1}' > output_data/
${peps_name}.rxlr_WYfold_hmm.list
# R environment
wy_list_re_allgenes <- "orfs_cand_RXLRs_.*add_rxlrs.
rxlr_WYfold_hmm.list"
allgenes_wys <- list.files(path = "output_data/",
 pattern = wy_list_re_allgenes,
 full.names = TRUE) %>%
 map_dfr(read_wy_list_allgenes) %>%
 unite("ID_isolate", ID, isolate, sep=":", remove =
FALSE) %>%
 mutate("method" = "WY_hmm")
```

4. Obtain a list of all secreted proteins that meet SignalP and TMHMM requirements.

```
# R environment
# SignalP results are to check if TM is before | after
the SP
sp3_fs_re_allgenes <- "*_longnames.str_stp.re-IDs.
add_rxlrs.rxlr_signalpep.outsp3_tabular.tmp"
sp3_fs_allgenes <- list.files(path = "output_data",
pattern = sp3_fs_re_allgenes, full.names = TRUE)
tmm_fs_re_allgenes <- ".*orfs_cand_RXLRs_.
*add_rxlrs.tmhmm.out"
tmm_fs_allgenes <- list.files(path= "output_data",
 pattern = tmm_fs_re_allgenes, full.names = TRUE)
# Dataframe for mapping, make sure filenames line up
between SP and TM
sp3_tmm_fs_allgenes <- tibble("sp3_fs" = sp3_fs_all-
genes,
 "tmm_fs" = tmm_fs_allgenes) %>%
 mutate("isolate_s" = str_extract(sp3_fs,
 "(?<=/orfs_cand_RXLRs_).*(?=_longnames)")) %>%
 mutate("isolate_t" = str_extract(tmm_fs,
 "(?<=/orfs_cand_RXLRs_).*(?=_longnames)"))
```

```
# Perform filtering function
sp3_tmm_pass_allgenes <- pmap(sp3_tmm_fs_allgenes, ~
read_prune_tms(..1, ..2, ..3),
 .id = 'isolate') %>%
 bind_rows() %>%
  unite("ID_isolate", Protein, isolate, sep=":",
remove = FALSE)
```

5. Scan for RxLR and EER domains.

```
# R environment
# Sequence filenames
seqs_fs_re_allgenes <- "orfs_cand_RXLRs_.*_long-
names.str_stp.re-IDs.add_rxlrs.fasta"
seqs_fs_allgenes <- list.files(path = "output_data",
 pattern = seqs_fs_re_allgenes,
 full.names = TRUE) %>%
 tibble("seqs_fs" = .) %>%
 separate(seqs_fs, into = c("isolate", NA),
 sep = "(?=_longnames)", remove = FALSE) %>%
  mutate(isolate = str_remove(isolate, ".*orfs_-
cand_RXLRs_"))
# SignalP filenames and join to seq fnames
seqs_sp3_fs_allgenes <- sp3_fs_allgenes %>%
 tibble("sp3_fs" = .) %>%
 separate(sp3_fs, into = c("isolate", NA),
 sep = "(?=_longnames)", remove = FALSE) %>%
  mutate(isolate = str_remove(isolate, ".*orfs_-
cand_RXLRs_")) %>%
 full_join(seqs_fs_allgenes, by = "isolate") %>%
 select(seqs_fs, sp3_fs, isolate)
# Run motif searches
allgenes_re_effectr <- map2(seqs_sp3_fs_allgenes
$seqs_fs, seqs_sp3_fs_allgenes$sp3_fs, effec-
tr_eer_sp3)
allgenes_re_effectr_tb <- tibble(bind_rows(allgen-
es_re_effectr)) %>%
 unite("ID_isolate", ID, isolate, sep=":", remove =
FALSE)
```

6. Filter down to final RxLR candidates based on motif patterns.

```
# R environment
allgenes_re_effectr_tb_meths <- allgenes_re_ef-
fectr_tb %>%
```

```
bind_rows(allgenes_wys) %>%
group_by(ID_isolate, ID, isolate) %>%
 summarize(methods_list = paste(method, collapse =
",")) %>%
filter((grepl("Whis2007", methods_list)) |
  (grepl("Whis_rxlr", methods_list) & (grepl("Whi-
s_eer", methods_list) | grepl("WY_hmm", methods_-
list))) |
 (grepl("Win2007", methods_list) & (grepl("Whis_eer",
methods_list) | grepl("WY_hmm", methods_list)))) %>%
mutate(isolate = str_remove(isolate, "/"))
```

7. Summarize final counts of RxLR candidates obtained.

```
# R environment
allgenes_re_effectr_tb_meths %>%
 filter(ID_isolate %in% sp3_tmm_pass_allgenes$ID_iso-
late) %>%
 group_by(isolate) %>%
 distinct(ID_isolate) %>%
   summarize(noTM_whis_or_whiswinrxlrEER_whiswinrxlr-
WY_count = n())
```

8. Write final list of candidates as a list. The filename used is based on the isolate ID used.

```
# R environment
outdir <- "output_data"
out_prefix <- "/allgenes_RXLRs_"
# Write each group as its own txt file
allgenes_re_effectr_tb_meths %>%
 filter(ID_isolate %in% sp3_tmm_pass_allgenes$ID_isolate)
%>%
 group_by(isolate) %>%
 distinct(ID_isolate, .keep_all = TRUE) %>%
 group_walk(~ writeLines(.x$ID,
 paste0(outdir, out_prefix, .y$isolate, ".txt")))
```

3.4 Notes

1. If required fields are absent from the provided GFF, some tools used in this protocol may add these features automatically. If this occurs, the tools will report their additions to the terminal. If gene annotations are available but not in GFF format, they can be converted into GFF format with tools such as AGAT.

2. To install and configure SignalP and TMHMM, read the README files that are included with the software downloads. The software may not work unless certain lines of the SignalP script are edited as described therein.
3. Unconventional secretion pathways have been documented in *Phytophthora* species. However, not requiring a signal peptide for RxLR genes results in excessive false positives. Since some Crinkler genes have been found without signal peptides [12], and since the "LFLAK" motif is likely part of an anciently conserved domain of secreted proteins that does not require a signal peptide [14], the presence of signal peptide and absence of TM helices are not required to classify Crinkler effectors.
4. To balance sensitivity to RxLR calls with specificity of not obtaining excessive false positives, multiple criteria are required. Candidates must have the RxLR motif as well as either an EER motif or match to a WY HMM profile.
5. As set up, the HMM search will only report hits with a bit score greater than 0 (-T 0). The bias filter implemented in hmmsearch may remove potentially real hits [15] and can be turned off by including—nobias.
6. Picking HMM score and SignalP version. Different versions of SignalP use different machine-learning algorithms and sequences for training data. Among early versions of SignalP, SignalP (v. 3.0) was found to be the most sensitive to classifying *Phytophthora* RxLR proteins. Sensitivities of SignalP versions 5 and 6 have not been compared in publications yet, but appear to be similarly or less sensitive than SignalP (v. 3.0). The sensitivity of SignalP (v. 4.0) can be adjusted to obtain similar results to SignalP (v. 3.0) [16].
7. If all that is required is a list of ORFs that are candidates to be RxLR effectors, this final step of Subheading 3.2 does not need to be performed. However, the locations of the candidates are often useful in combination with expression data or to analyze genome compartmentalization. The GFF can easily be converted to other useful formats such as BED with other tools, e.g., gff2bed from BEDOPS.
8. Adding start and stop codons is useful to validate that all code was executed correctly and that the inferred positions of all candidate ORFs match the correct parts of the genome assembly. All candidates should have start codons, but some may not contain stop codons. Although EMBOSS getorf was parameterized in **step 1** to only output sequences between start and stop codons, it will also output a fragmented ORF if the contig end is reached before a stop codon. When uploading to a genome database such as NCBI, fragmented genes may cause warnings, but these are legitimate effector candidates that may

have been captured with a more complete or contiguous assembly.

9. Expected protocol outputs are listed below.

<assembly_name > .orfs-min70long.start2stop.fasta: This file contains FASTA sequences of open reading frames (ORFs) from the genome assembly. The ORFs are at least 70 amino acids long and are derived from the start codon to the stop codon.

<assembly_name > .orfs-min70long.start2stop.rxlr_signalpepsp3_tabular.tmp: This is a tab-delimited file that contains prediction scores for various features of the ORFs from the rxlr_signalpepsp3 model. The model predicts various scores like Cmax_score, Ymax_score, Smax_score, etc., and a prediction (Y or N) based on these scores.

<assembly_name > .orfs-min70long.start2stop.rxlr_WY-fold_hmm.list: This file contains a list of identifiers for ORFs that likely contain the WY-fold domain, as identified by a Hidden Markov Model (HMM) analysis.

<assembly_name > .orfs-min70long.start2stop.rxlr_WY-fold_hmm.log: This is a log file generated during the HMMER analysis which searched for the WY-fold domain in the ORFs.

<assembly_name > .orfs-min70long.start2stop.rxlr_WY-fold_hmm.out: This file contains the output of the HMMER analysis, which includes details such as e-value, score, and bias for each target sequence.

<assembly_name > .orfs-min70long.start2stop.tmhmm.out: This file contains the output of the TMHMM analysis, which predicts transmembrane helices in proteins.

orfs_cand_RXLRs_ < assembly_name > _longnames.gff: This is a General Feature Format (GFF) file that contains the genomic coordinates of the candidate RxLR effector genes.

orfs_cand_RXLRs_ < assembly_name > _assembly_longnames.list: This file contains a list of candidate RxLR effector genes with their genomic coordinates.

orfs_cand_RXLRs_ < assembly_name > _longnames.merge_ovlp.gff: This is a modified GFF file after merging overlapping ORFs.

orfs_cand_RXLRs_ < assembly_name > _longnames.str_stp.gff: This is a GFF file with start and stop codons of the ORFs.

orfs_cand_RXLRs_ < assembly_name > _longnames.str_stp.re-IDs.add_rxlrs.fasta: This is a FASTA file containing sequences of the candidate RxLR effector genes.

orfs_cand_RXLRs_ < assembly_name > _longnames.str_stp.re-IDs.add_rxlrs.gff: This is a GFF file containing the genomic coordinates of the candidate RxLR effector genes with re-annotated identifiers.

orfs_cand_RXLRs_ < assembly_name > _longnames.str_stp.re-IDs.add_rxlrs.rxlr_signalpepsp3_ tabular.tmp: This is a tab-delimited file that contains prediction scores for various features of coding genes from the rxlr_signalpepsp3 model. The model predicts various scores like Cmax_score, Ymax_score, Smax_score, etc., and a prediction (Y or N) based on these scores.

orfs_cand_RXLRs_ < assembly_name > _longnames.str_stp.re-IDs.add_rxlrs.rxlr_WYfold_hmm. list: This file contains a list of identifiers for coding genes that likely contain the WY-fold domain, as identified by a Hidden Markov Model (HMM) analysis.

orfs_cand_RXLRs_ < assembly_name > _longnames.str_stp.re-IDs.add_rxlrs.rxlr_WYfold_hmm. log: This is a log file generated during the HMMER analysis which searched for the WY-fold domain in the coding genes.

orfs_cand_RXLRs_ < assembly_name > _longnames.str_stp.re-IDs.add_rxlrs.rxlr_WYfold_hmm. out: This file contains the output of the HMMER analysis, which includes details such as e-value, score, and bias for each target sequence.

allgenes_RXLRs_ < assembly_name > .txt: This file contains a list of all candidate RxLR effector genes identified in the genome assembly.

10. The output from this step includes the final GFF with RxLR candidates added. All candidates will be at the bottom of the GFF. The companion Python script can be parameterized to sort the effectors by position, together with other genes. However, it is slow and, depending on feature types in the GFF, the sorting may be unpredictable. For sorting that keeps compliance with specs for other tools such as bedops, parse the GFF through any script in AGAT. Some features may be added or their positions adjusted, and AGAT reports all such changes to the terminal.

Acknowledgements

This work was supported by grants from USDA ARS (2072-22000-041-000-D), USDA NIFA (2023-67013-39918) and USDA APHIS to N.J.G. and a USDA ARS grant (2021-67034-35171) to N.C.

References

1. Jones J, Dangl J (2006) The plant immune system. Nature 444:323–329. https://doi.org/10.1038/nature05286
2. Whisson SC, Boevink PC, Moleleki L et al (2007) A translocation signal for delivery of oomycete effector proteins into host plant cells. Nature 450:115. https://doi.org/10.1038/nature06203
3. Rehmany AP, Gordon A, Rose LE et al (2005) Differential recognition of highly divergent downy mildew avirulence gene alleles by RPP1 resistance genes from two *Arabidopsis* lines. Plant Cell 17:1839–1850. https://doi.org/10.1105/tpc.105.031807
4. Tyler BM, Putnam NH, Phuntumart V et al (2006) *Phytophthora* genome sequences uncover evolutionary origins and mechanisms of pathogenesis. Science 313:1261–1266. https://doi.org/10.1126/science.1128796
5. Kronmiller BA, Feau N, Shen D et al (2023) Comparative genomic analysis of 31 *Phytophthora* genomes reveals genome plasticity and horizontal gene transfer. MPMI 36:26–46. https://doi.org/10.1094/MPMI-06-22-0133-R
6. Lévesque CA, Brouwer H, Cano L et al (2010) Genome sequence of the necrotrophic plant pathogen *Pythium ultimum* reveals original pathogenicity mechanisms and effector repertoire. Genome Biol 11:R73. https://doi.org/10.1186/gb-2010-11-7-r73
7. Adhikari BN, Hamilton JP, Zerillo MM et al (2013) Comparative genomics reveals insight into virulence strategies of plant pathogenic oomycetes. PLoS One 8:e75072. https://doi.org/10.1371/journal.pone.0075072
8. Ai G, Yang K, Ye W et al (2020) Prediction and characterization of RXLR effectors in *Pythium* species. MPMI 33:1046–1058. https://doi.org/10.1094/MPMI-01-20-0010-R
9. Tabima JF, Grünwald NJ (2019) *effectR* : an expandable R package to predict candidate RxLR and CRN effectors in oomycetes using motif searches. MPMI 32:1067–1076. https://doi.org/10.1094/MPMI-10-18-0279-TA
10. Charif D, Lobry JR (2007) SeqinR 1.0-2: a contributed package to the R project for statistical computing devoted to biological sequences retrieval and analysis. In: Bastolla U, Porto M, Roman HE, Vendruscolo M (eds) Structural approaches to sequence evolution. Springer, Berlin, Heidelberg, pp 207–232
11. Sperschneider J, Williams AH, Hane JK et al (2015) Evaluation of secretion prediction highlights differing approaches needed for oomycete and fungal effectors. Front Plant Sci 6. https://doi.org/10.3389/fpls.2015.01168
12. Haas BJ, Kamoun S, Zody MC et al (2009) Genome sequence and analysis of the Irish potato famine pathogen *Phytophthora infestans.* Nature 461:393–398. https://doi.org/10.1038/nature08358
13. Win J, Morgan W, Bos J et al (2007) Adaptive evolution has targeted the C-terminal domain of the RXLR effectors of plant pathogenic oomycetes. Plant Cell 19:2349–2369. https://doi.org/10.1105/tpc.107.051037
14. Zhang D, Burroughs AM, Vidal ND et al (2016) Transposons to toxins: the provenance, architecture and diversification of a widespread class of eukaryotic effectors. Nucl Acids Res 44:3513–3533. https://doi.org/10.1093/nar/gkw221
15. Nur M, Wood K, Michelmore R (2023) EffectorO: motif-independent prediction of effectors in oomycete genomes using machine learning and lineage specificity. Mol Plant-Microbe Interact 36:397. https://doi.org/10.1094/MPMI-11-22-0236-TA
16. Wood KJ, Nur M, Gil J et al (2020) Effector prediction and characterization in the oomycete pathogen *Bremia lactucae* reveal host-recognized WY domain proteins that lack the canonical RXLR motif. PLoS Pathog 16:e1009012. https://doi.org/10.1371/journal.ppat.1009012

Chapter 12

Pinus radiata cDNA Library for the Screening of *Phytophthora* Effector Protein Targets in Yeast

Sophie Eccersall and Claudia-Nicole Meisrimler

Abstract

Yeast two-hybrid library screening enables the discovery of novel protein-protein interactions. Identifying cytosolic host proteins targeted by host-translocated *Phytophthora* effector proteins relies on the mRNA amount, quality, and composition used to prepare the yeast two-hybrid cDNA library. Here we describe the steps required for the preparation of a *Pinus radiata* cDNA library optimized for *Phytophthora* effector target screening in yeast. This method may also be applied to other host plants and pathogen species.

Key words Yeast two-hybrid, mRNA purification, cDNA library, *Escherichia coli*, Yeast

1 Introduction

In 1989, Fields and Song developed the two-hybrid technique in yeast as a protein interaction detection technology [1]. Since then, many laboratories have used a variety of protocols that enable cDNA and genomic fusion libraries to be screened for protein interaction partners [1–5]. In the last decade, yeast two-hybrid (Y2H) libraries have become an important tool to identify novel effector protein targets in host plants and other protein- and peptide-protein interactions in the host and pathogen [6–10].

A Y2H library screening is based on the activation of downstream reporter gene(s) by the binding of a transcription factor onto an upstream activating sequence. Transcription factors used in Y2H are split into two functional domains: the DNA-binding domain and the transcription-activation domain. Importantly, the DNA-binding domain is responsible for binding to the upstream activating sequence, whereas the transcription activation domain activates the transcription [11]. In Y2H, each of the domains is artificially fused to a protein. Respectively, when these proteins interact, the DNA-binding domain and transcription activation domain form a functional transcription factor, allowing activation

Monica L. Gerth and Rosie E. Bradshaw (eds.), *Phytophthora: Methods and Protocols*, Methods in Molecular Biology, vol. 2892, https://doi.org/10.1007/978-1-0716-4330-3_12,

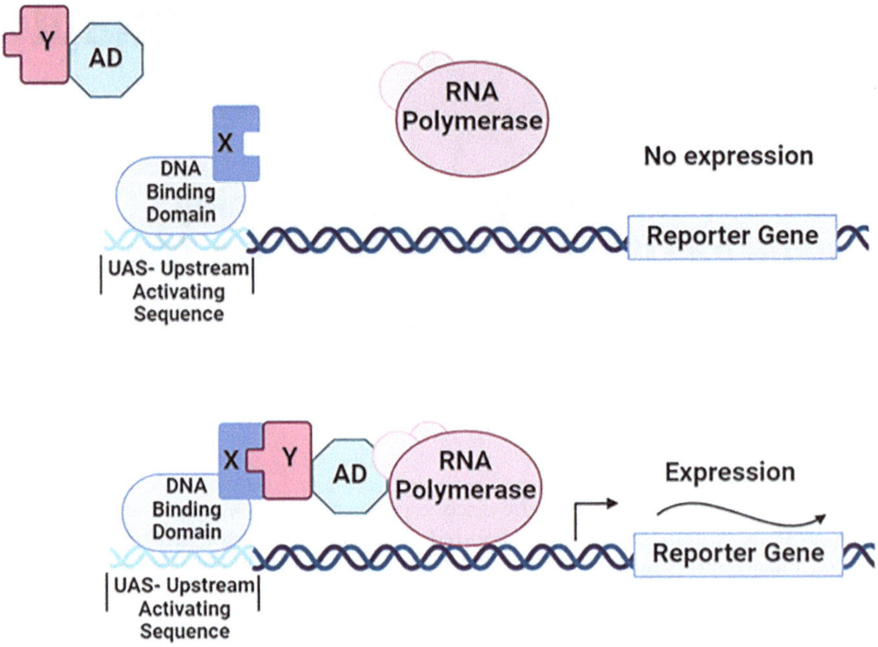

Fig. 1 Overview of the principle underpinning yeast two-hybrid (Y2H) systems. The basic concept of the Y2H system is to detect the interaction between two proteins via transcriptional activation of a reporter gene. Y2H systems exploit the fact that transcription factors comprise two domains, a DNA binding domain (DBD) and an activation domain (AD). Two separate hybrid proteins are constructed in two-hybrid screens. The first hybrid protein is the DBD/protein X fusion, known as the "bait," while the second hybrid protein is the AD/protein Y fusion known as the "prey." These two hybrids are encoded on separate yeast expression plasmids, with independent selectable markers. If the bait interacts with the prey in the nucleus, this will bring the activation domain together with the DNA-binding domain to reconstitute transcriptional activation and result in the expression of the reporter gene(s)

of downstream reporter gene(s) by the binding onto the upstream activating sequence. Hence, the Y2H is a protein-fragment complementation assay comparable to Split-YFP or Split-Ubiquitin assays [12, 13]. Ultimately, the activation of downstream reporter genes allows visualization of the interaction and enables identification of the interacting molecules.

For the purposes of identifying new *Phytophthora* effector protein targets in plants, cDNA encoding the effector protein (bait protein) is ligated into a yeast shuttle vector (e.g., pDEST32™) to create a fusion protein with the DNA binding domain of the GAL4 transcription factor (*see* Fig. 1). To test these bait proteins, a library of potential plant target proteins (prey), each fused with the activation domain of the GAL4 transcription factor, is also generated. The primary focus of this chapter is the preparation of a prey library. The example we present here is a prey library preparation using the CloneMiner™ II cDNA Library Construction Kit [18]. This example uses tissues from different *Pinus radiata* needle material infected and treated with immunity-related hormones. *Pinus radiata* needles, together with Eucalyptus and Oak leaves, form a class of extremely recalcitrant samples; nevertheless, they are important hosts for a wide range of *Phytophthora* species. Ultimately, the cDNA prey library is ligated into a second shuttle vector (pDEST22™) to create an activation domain-tagged cDNA library (*see* Fig. 1), which can be used for a Y2H screening either using co-transformation in a single yeast (e.g., MaV203) or mating of two compatible yeast strains (e.g., Y8800, Y8930) [6, 14, 15].

2 Materials

Prepare all solutions using sterile ultrapure water (prepared by autoclaving deionized water, purified to a sensitivity of 18 MΩ-cm at 25 °C) and ultrapure grade reagents. Prepare and store all reagents at room temperature under sterile conditions. In order to obtain prey cDNA that includes the host plant defence proteins, infected plant material with the same pathogen is used to make the bait protein. Additional chemical treatments can also be used to stimulate immune responses in the host plant.

2.1 Bait Preparation

1. cDNA originating from the organism of interest, such as a *Phytophthora* species (*see* **Note 1**).
2. A suitable GAL4 DNA Binding Domain containing vector, e.g., pDEST32™ (~200 ng/μL), purified using a miniprep plasmid extraction kit (*see* **Note 2**).

2.2 Plant Material

1. *Pinus radiata* clonal rootless pine shoots (also called pine plantlets) produced by adventitious shoot production from cotyledons. Alternatively, fresh pine needles collected on the day of the procedure from a tree of at least 1 year of age can be used.
2. 50 mL tubes.
3. 10 mM benzothiadiazole (BTH).
4. 500 μM flagellin 22.
5. 1 mM jasmonic acid.
6. 10 mM salicylic acid.
7. Zoospore suspension from the bait organism of choice (*see* **Notes 3** and **4**).
8. Aerial spray machine or airbrush.
9. Petri dishes.
10. Parafilm or similar tape.
11. Climate-controlled incubator with light (e.g., Contherm Biosyn Incubator 6000CP).

2.3 RNA Extraction and mRNA Purification

1. Plant total RNA extraction kit, e.g., RNeasy PowerPlant Kit (Qiagen) (*see* **Note 5**).
2. High-velocity bead beater.
3. RNase decontamination solution.
4. Mortar and pestle capable of tolerating liquid nitrogen.
5. Liquid nitrogen.
6. 100% ethanol.
7. β-Mercaptoethanol.
8. RNA loading gel dye.
9. NanoDrop One Microvolume UV-Vis Spectrophotometer or similar.
10. mRNA Purification Kit, e.g., Dynabeads™ (ThermoFisher Scientific).
11. Neodymium magnet rack.

2.4 Quality Control for the cDNA Library Using Restriction Digestion

1. Appropriate restriction enzyme for the vector of choice, e.g., *BsrGI.*
2. 1 KB Plus DNA ladder.
3. Electrophoresis apparatus.
4. Tris-acetate EDTA buffer (TAE; 50×): 2.0 M Tris base, 1.0 M acetic acid, 0.05 M EDTA, pH 8.0. Weigh 242 g Tris base and transfer to a 2 L beaker. Add 750 mL water and stir until tris base has completely dissolved. Add 100 mL of 0.5 M EDTA solution and 57.1 mL glacial acetic acid. Allow solution to mix; adjust pH if necessary.
5. 1 × TAE buffer: 40 mM Tris base, 20 mM acetic acid, 1 mM EDTA. Take 1 volume of concentrated stock solution and add 49 volumes of distilled water, mix thoroughly.
6. Miller's Luria-Bertani (LB) broth base (Thermo Fisher Scientific).
7. Kanamycin (50 mg/mL).
8. Plasmid DNA miniprep kit, e.g., Macherey-Nagel™ NucleoSpin Plasmid QuickPure™ Kit (Macherey-Nagel).

2.5 cDNA Library for Yeast-2-Hybrid

1. mRNA extracted from infected plant needles (for example: Control, 3-DAI (days after infection), 5-DAI, 7-DAI).
2. CloneMiner™ II cDNA Library Construction Kit (Thermo Fisher Scientific).
3. 0.5 M EDTA, pH 8.0.
4. 160 uL phenol: chloroform: isoamyl alcohol (25:24:1).
5. 7.5 M ammonium acetate (NH_4OAc).

6. 100% ethanol.
7. 70% ethanol.
8. Tris EDTA NaCl (TEN) buffer (10 mM Tris-HCl, pH 7.5; 0.1 mM EDTA; 25 mM NaCl).
9. Tris EDTA (TE) buffer (10 mM Tris-HCl, pH 8.0; 1 mM EDTA).
10. SOC medium (2% tryptone, 0.5% yeast extract, 10 mM NaCl, 2.5 mM KCl, 10 mM $MgCl_2$, 10 mM $MgSO_4$, 20 mM glucose).
11. Freezing media (60% SOC medium: 40% glycerol).
12. Miller's Luria-Bertani (LB) broth base (Thermo Fisher Scientific).
13. LB agar plates.
14. Glycogen (20 μg/μL).
15. Ampicillin stock (100 mg/mL).
16. Gentamicin stock (30 mg/mL).
17. Kanamycin stock (50 mg/mL).
18. PCR thermocycler or heat block.

2.6 Cloning and Transformation

1. pDONR221 (alternative pDONR222), pDEST32™, pDEST22™ (Thermo Fisher Scientific).
2. Plasmid DNA miniprep kit, e.g., Macherey-Nagel™ NucleoSpin Plasmid QuickPure™ Kit (Macherey-Nagel).
3. Plasmid DNA midiprep kit, e.g., PureLink™ HiPure Plasmid Filter Midiprep Kit (Thermo Fisher Scientific).
4. Gateway® BP Clonase® II enzyme mix (keep at −20 °C until immediately before use) (Thermo Fisher Scientific).
5. Gateway® LR Clonase® II enzyme mix (Thermo Fisher Scientific).
6. SOC medium (*see* Subheading 2.5).

3 Methods

Conduct all procedures in strict sterile conditions at room temperature unless otherwise specified.

3.1 Bait Preparation

Generating your bait protein from your pathogen allows for the testing of pairwise interactions and library screening against the prey cDNA library.

1. Generate bait of your protein of interest via cloning into an appropriate destination vector (e.g., pDEST32™) (*see* **Notes 1** and **2**).

3.2 Collection and Treatment of Pine Needles

The plant tissues form the basis of the cDNA library. It is at this step you may change which plant tissue and pathogen you wish to test and what conditions, treatments, and time points you select for use. In the example given here, both pathogen infection and chemical compounds are used to induce a broad range of immune responses in a pine host plant, the transcripts of which we wish to capture.

1. Collect fresh pine needles at the start of the experiment.
2. For pathogen infection, take 5 mL of zoospore solution and gently pipette into 50 mL centrifuge tubes (*see* **Notes 3** and **4**). Place pine needles fascicle end down into the zoospore solution.
3. For sample treatment with chemical compounds, spray with solutions (100 μM jasmonic acid, 100 μM BTH, 100 μM salicylic acid, and 1 μM Flg22) thoroughly to ensure even coverage.
4. Cap the tubes and cover in foil, then leave overnight (18–24 h) standing upright.
5. Flash freeze chemically treated needles in liquid nitrogen and store at −80 °C until needed.
6. For pathogen infection samples, prepare Petri dishes by layering three to four sterilized paper towels into each dish. Thoroughly dampen the paper towels with sterile water.
7. Place needles incubated with spores as flat as possible into the dishes and seal the dishes with parafilm.
8. Place into an incubator (for *Phytophthora pluvialis,* set the temperature at 17 °C; 12 h daylight, 12 h darkness), and check at 1-DAI (day after infection), 2-DAI, 3-DAI, 5-DAI and 7-DAI intervals (*see* **Note 5**).
9. At the appropriate time points, you have the option of (a) immediately performing an RNA extraction on the needles, or (b) flash freezing the samples in liquid nitrogen and storing at −80 °C (*see* **Note 6**).

3.3 Total RNA Extraction and mRNA Purification

RNA extraction is integral to the creation of a cDNA library, as it forms the template from which the cDNA will be created. Various extraction kits exist, with many brands having kits designed for specific types of tissues. It is recommended using a kit or method which allows for maximum extraction from your tissue type.

1. Before starting, treat the work area with RNase decontamination solution (*see* **Note 7**).

2. Thoroughly macerate ~50 mg pine tissue in a mortar and pestle with liquid nitrogen until the tissue is powdered (*see* **Note 8**).
3. Perform RNA extraction according to kit instructions (*see* **Notes 9–14**).
4. To elute RNA, add 25 μL of appropriate RNA elution solution to the column. Centrifuge for 1 min at 200× *g*, then for 2 min at 20,000× *g*.
5. Repeat **step 4** to ensure all RNA is eluted.
6. Keeping eluted RNA on ice, measure concentration with a spectrophotometer.
7. Store eluted RNA at −80 °C until ready for further use or move immediately onto mRNA purification.

3.4 mRNA Purification

While only a small percentage of a cells' total RNA (1–5%), mRNA represents all the transcripts in the host cell at the time of isolation. In eukaryotes, mRNA contains a poly-(A) tail, which can be used as a primer site for reverse transcription. The purified mRNA extracted from this step provides a template for the synthesis of cDNA.

1. Using an mRNA isolation kit (e.g., ThermoFisher Scientific Dynabeads™), purify the mRNA from the total RNA as per the manufacturer's instructions and store at −80 °C until ready for further use. You will need 5 μg total mRNA to construct a standard library (*see* **Notes 15** and **16**).

3.5 cDNA Library

This step creates the cDNA library from the purified mRNA. By creating a cDNA library, different bait proteins of interest can be screened for interactions. From this, potential targets of these baits can be identified. Fractionation columns are used to generate cDNA that is free from contaminating elements such as small DNA and adapters (*see* **Note 17**).

1. Dilute the purified mRNA to 5 ug in a total volume of 7 uL with DEPC treated water.
2. Add 1 μL of Biotin-attB2-Oligo(dT) Primer (CloneMiner™ II cDNA Library Construction Kit; Thermo Fisher Scientific) (*see* **Notes 18** and **19)** to the diluted mRNA and gently mix, then centrifuge briefly (i.e., 2 s) to collect the sample.
3. Incubate the mixture at 70 °C in a heat block for 7 min and allow it to gradually cool to 45 °C over 30 min.
4. While the mixture is incubating, prepare the reagents for synthesizing the first cDNA strand. To a fresh microcentrifuge tube, add the following kit reagents: 4 μL of 5× First Strand Buffer, 2 μL of 0.1 M DTT, 1 μL of 10 mM dNTPs (*see* **Note 18**). Mix gently and place on ice.

5. After the mixture from **step 3** has cooled to 45 °C, add the mixture from **step 4**. Gently pipette them together, being careful not to introduce bubbles.
6. Incubate the mixture at 45 °C for two min.
7. While the mixture is incubating, carefully add 5 uL of Reverse Transcriptase (e.g., SuperScript® III RT) and gently mix.
8. Incubate the reaction tube in stepwise increments: 45 °C for 20 min; 50 °C for 20 min; 55 °C for 20 min. Place the tube on ice.
9. Keep the following reagents on ice and work quickly. To the tube from **step 8** (*see* **Note 18**), add:
   - 91 μL DEPC-treated water
   - 30 μL 5× Second Strand Buffer
   - 3 μL 10 mM (each) dNTPs
   - 1 μL *Escherichia coli* (*E. coli*) DNA Ligase (10 U/μL)
   - 4 μL *E. coli* DNA Polymerase I (10 U/μL)
   - 1 μL *E. coli* RNase H (2 U/μL)
10. Gently mix the contents together and briefly centrifuge to collect the sample.
11. Incubate the tube at 16 °C for 2 h.
12. Add 2 μL of T4 DNA Polymerase (*see* **Note 18**); mix gently by pipetting and briefly centrifuge to collect the sample.
13. Incubate at 16 °C for 5 min.
14. Add 10 μL 0.5 M EDTA, pH 8.0.
15. Add 160 μL of phenol: chloroform: isoamyl alcohol (25:24:1) and mix by hand for 30 s.
16. Centrifuge at room temperature for 5 min at 16,000× *g*. Carefully remove the upper aqueous phase and transfer it to a fresh 1.5-mL tube. Avoid transferring any phenol during pipetting.
17. To the collected aqueous phase, add 1 μL glycogen (20 μg/μL), 80 μL 7.5 M NH_4OAc, and 600 μL 100% ethanol.
18. Place the tube at −80 °C for 1 h. You may stop at this point if necessary and store the mixture at −20 °C overnight instead.
19. Centrifuge the sample at 4 °C for 35 min at 16,000× *g*.
20. Remove the supernatant while trying not to disturb the cDNA pellet. Add 150 μL of 70% ethanol.
21. Centrifuge the sample at 4 °C for 2 min at 16,000× *g*. Carefully remove the supernatant.
22. Repeat the 70% ethanol wash and centrifugation. Remove as much of the remaining ethanol as possible.

23. Allow the cDNA pellet to dry at room temperature for 30 min or until dry.
24. Resuspend the pellet in 22 μL DEPC-treated water by pipetting up and down.
25. Centrifuge briefly to collect the sample and place it on ice.
26. Keep the cDNA on ice and add the following (*see* **Notes 18** and **19**):
    - 10 μL Standard 5× Adapter Buffer
    - 4 μL attB1 Adapter (1 μg/μL)
    - 8 μL 0.1 M DTT
    - 6 μL T4 DNA Ligase (1 U/μL)
27. Mix the contents gently by pipetting. Incubate at 16 °C for 16–24 h.
28. Incubate the reaction tube at 70 °C for 10 min.
29. Keep the reaction tube on ice while you set up the fractionation column (*see* **Note 18**). Fractionation removes smaller DNA fragments, as well as adapters, primers, and non-recombinant clones.
30. Secure the column in a support stand and place a rack underneath containing one 15 mL tube.
31. The fractionation columns provided are prepacked with 20% ethanol. Remove the top cap first, then the bottom cap, allowing the ethanol to drain out of the column by gravity into the 15 mL tube.
32. Once the column stops dripping, pipette 800 μL of TEN buffer (*see* **Note 18**) into the column and allow it to drain completely. Repeat three more times (four washes total). Allow the column to drain until dry (*see* **Note 20**).
33. Label three sterile 1.5 mL tubes 1–3 and place into the rack. *The rack needs to be able to freely move beneath the columns; the outlet of the column should be 1–2 cm above the tubes.*
34. Place Tube 1 underneath the column.
35. Add 100 μL of TEN buffer to the cDNA reaction tube sitting on ice. Gently mix and briefly centrifuge to collect the sample.
36. Add the entire sample to the column; collect the eluate into Tube 1.
37. Move Tube 2 beneath the column and add 240 μL of TEN buffer to the column. Let the column drain completely and collect the eluate into tube 2. The majority of the cDNA will be collected into this tube.
38. Move Tube 3 beneath the column and add 80 μL of TEN buffer to the column. Let the column drain completely and collect the eluate into tube 3.

39. Check the cDNA quantity that has been eluted into the tubes via a spectrophotometer; store Tube 3 in −20 °C in case it is required later (*see* **Note 21**).
40. To tube 2, add 1 μL of glycogen (20 ug/μL), 120 μL of 7.5 M NH_4OAc and 900 μL of 100% ethanol.
41. Place the tube at −80 °C for 1 h to precipitate the cDNA.
42. Centrifuge tube at 4 °C for 35 min at 16,000× *g*.
43. Carefully remove the supernatant; add 150 μL of 70% ethanol to the cDNA pellet.
44. Centrifuge the sample at 4 °C for 2 min at 16,000× *g*. Remove the supernatant; repeat the ethanol wash and centrifugation step then remove as much of the ethanol as possible and allow the cDNA pellet to dry at room temperature for 45 min or until dry.
45. Resuspend the cDNA pellet in 7 μL of TE buffer by pipette; transfer the sample to a new tube.
46. Check the yield of the cDNA using a spectrophotometer (*see* **Note 21**).

3.6 cDNA Transfer

Once the cDNA has been purified, it will be cloned into the appropriate vector to create the library.

1. To a fresh microcentrifuge tube, add 4 μL of the cDNA, 2 μL of pDONR221 plasmid (150–200 ng/μL) and 1 μL of TE buffer (*see* **Note 22**).
2. Add 3 μL of BP clonase and gently mix by pipetting. Briefly centrifuge to collect the samples.
3. Prepare positive and negative controls (*see* **Notes 18** and **23**) as follows:

Component	Negative BP control (μL)	Positive BP control (μL)
pDONR221 (150–200 ng/μL)	1.67	1.67
pEXP7-tet positive control	–	0.5
TE buffer, pH 8.0	To 7	To 7
BP Clonase	3	3

4. Incubate the reactions for 16–20 h at room temperature.
5. The following day, add 2 μL of Proteinase K to each reaction tube and incubate at 37 °C for 15 min, then 75 °C for 10 min.
6. To each BP reaction, add sterile water (NOT DEPC TREATED) to make a total volume of 100 μL.

7. Add 1 μL of glycogen (20 ug/μL), 50 μL of NH_4OAc 7.5 M and 375 μL of 100% ethanol to each reaction. Place the tubes at −80 °C for 1 h; centrifuge the samples at 4 °C for 35 min at 16,000× *g*.
8. Remove the supernatant carefully and add 150 μL of 70% ethanol to the pellet. Centrifuge the samples at 4 °C for 2 min at 16,000× *g*.
9. Repeat **step 8**. Remove the remaining ethanol and dry the cDNA pellet at room temperature for 45 min, or until dry.
10. Resuspend the pellet in 9 μL of TE buffer by pipetting.
11. Aliquot the cDNA library into six 1.5 mL tubes containing 1.5 μL each, and place tubes on ice.
12. Thaw ElectroMAX DH10B competent cells on ice. Other alternative electrocompetent cells lacking the F′ episome can be used as well (*see* **Note 24**).
13. To each sample, add 50 μL of the competent cells and gently pipette up and down to mix.
14. Prepare a pUC19 control by adding 1 μL of the pUC19 plasmid to 50 μL of competent cells.
15. Transfer each sample into an ice-cold cuvette. Gently tap the sides to distribute the contents evenly. Ensure there are no air bubbles.
16. Starting with the control samples, electroporate the samples at the optimized settings for your electroporation machine.
17. Add 1 mL of SOC medium to each sample and gently mix by inverting the cuvette until the samples and SOC medium appear well mixed. LB can be used as an alternative to SOC medium.
18. Gently tip each sample into separate 15 mL tubes. Place tubes on their sides and shake cells for 2 h at 37 °C at 225–250 rpm.
19. Following incubation, pool all the cDNA samples into one 15 mL tube.
20. Take 200 μL from each sample and place in 1.5 mL tubes for titre determination.
21. Add an equal amount of sterile freezing media to the remaining cDNA and control samples and store at −80 °C.

3.7 Analyzing the cDNA Library in *E. coli*

After creating the cDNA library, it is important to verify the titre and quality of the library. A plating assay will determine the titre of the library (e.g., how many primary clones are present), and a restriction digest is used to determine the average insert size and percentage of recombinants in the library.

1. Prepare two prewarmed LB plates containing ampicillin (100 ug/μL) for the pUC19 control sample.
2. Prepare 18 prewarmed LB plates containing kanamycin (50 ug/μL). You will be plating in duplicate and will need six plates for each cDNA library and for each control.
3. Serially dilute the samples with SOC medium following the table below. For each dilution, take 100 μL of sample and add to 900 μL of SOC medium.

	cDNA library	BP negative control	BP positive control	pUC19 control
Dilutions	10^{-2} 10^{-3} 10^{-4}	Undiluted 10^{-1} 10^{-2}	10^{-2} 10^{-3} 10^{-4}	10^{-2} - -
Amount on each plate	100 μL	100 μL	100 μL	100 μL
Total no° of kan plates	6	6	6	–
Total no° of amp plates	–	–	–	6

4. Spread 100 μL of each dilution onto the appropriate prewarmed plates and seal the plates with parafilm.
5. Incubate overnight at 37 °C.
6. Use the following equation to calculate the titre for each plate:

   cfu/mL = (colonies on plate × dilution factor)/volume plated,

   Calculate the average titre for the cDNA library.
7. Use the following equation to determine the total number of colony-forming units:

   Total cfu = average titre (cfu/mL) × total volume of cDNA library (mL).

   A well-represented library should contain 5×10^6 to 1×10^7 primary clones (*see* **Note 25**).
8. Take 24 colonies from the plating assay following the cDNA library construction and culture overnight in 5–10 mL of LB broth containing 50 μg/mL of kanamycin.
9. Isolate plasmids following your preferred manufacturer's plasmid purification kit.
10. Digest 300–500 ng of plasmid DNA with *BsrG* I following the manufacturer's instructions. Additionally, digest ~250 ng of pDONR 221 as a control (*see* Fig. 2) to determine the average insert size and percentage of recombinants (*see* **Note 26**).

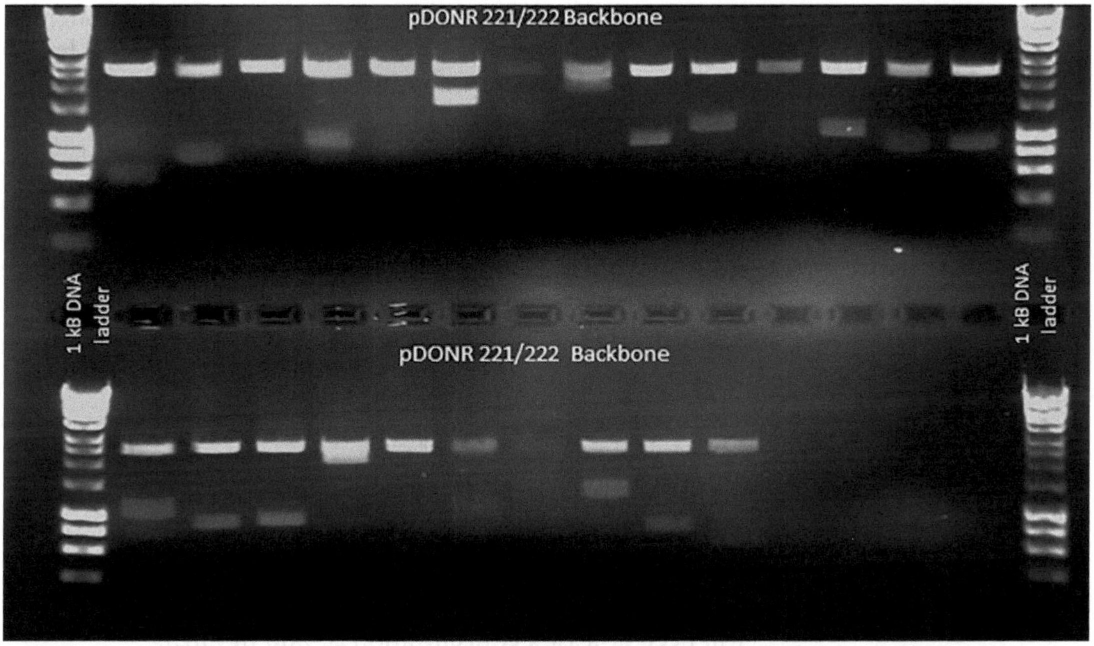

Fig. 2 BsrG I restriction digestion of the cDNA library in pDONR221/pDONR222. Twenty-four colonies were randomly chosen from the cDNA library (i.e., *Escherichia coli* transformed with pDONR221/pDONR222) to check for inserts. BsrG I was used for restriction digestion, followed by electrophoresis (TAE buffer, 1% agarose gel) of the products. SYBR™ Safe DNA Gel Stain was used in the samples for visualization of the DNA. A 1 kb DNA ladder was used as reference (lanes 1, 16)

11. The pDONR™ 221 control should show a digestion pattern of 3 bands of the following lengths: ~2.5 kb, ~1.4 kb, ~790 bp (*see* **Note 27**).
12. Each cDNA entry clone should have a vector backbone band of 2.5 kb and additional insert bands of varying sizes.
13. Identify clones containing inserts using the guidelines outlined above. Determine the percentage of clones containing plasmids with inserts (percentage recombinants).
14. For clones containing inserts, use the DNA ladder to estimate band sizes. If multiple bands exist for a single cDNA entry clone, add all band sizes to calculate the insert size. Do not include the 2.5 kb vector backbone band in your calculations. Calculate the average insert size for your cDNA library.

3.8 LR Library Transfer Reaction

Once the cDNA library has been constructed, it requires transformation into the appropriate destination vector to create an expression library which will allow for functional analysis.

1. Prepare plasmid DNA (minimum of 100 ng of DNA) from the cDNA library using a Midiprep plasmid purification kit of your choice.

2. Following the manufacturer's instructions from the ClonerMiner II kit, perform a PEG (30% PEG 8000/30 mM $MgCl_2$) precipitation and determine the DNA yield via a spectrophotometer; dilute to 25 ng/μL.
3. Combine 2 μL of the cDNA entry library with 3 μL of destination vector (pDEST22™, 200 ng/μL) and 9 μL of TE buffer, pH 8.0, in a sterile microcentrifuge tube.
4. Add 6 μL of LR clonase to the microcentrifuge tube (*see* **Note 23**). Mix well by very briefly vortexing (~2 s).
5. Incubate the reaction mixture overnight (16–18 h) at room temperature.
6. The following day, follow the CloneMiner II kit instructions for ethanol precipitation of the LR reaction.
7. Transform the LR reaction into an appropriate *E. coli* strain, such as Electromax™ T1 Phage Resistant cells for maximum transformation efficiency (*see* **Note 24**).
8. Prepare plasmid DNA with your miniprep kit of choice (you will need at least a concentration of 200 ng/μL).

3.9 Transformation of Bait and Prey into Yeast for Y2H Screening

1. To transform your cDNA library into your yeast strain of choice, the authors recommend following McLellan, H., Armstrong, M.R., Birch, P.R.J. (2021). Yeast Two-Hybrid Screening for Identification of Protein-Protein Interactions in *Solanum tuberosum* [16].

4 Notes

1. cDNA of the protein from the organism of interest can be generated via reverse transcription from total RNA with your preferred reverse transcriptase enzyme. Alternatively, gene fragments can be synthesized commercially (e.g., gBlocks/gene fragments from IDT) and inserted into the pDEST32™ by Gateway cloning. For more information on Gateway cloning, see reference [17]. The gene of interest can also be synthesized directly into the pDEST32™ vector.
2. If you plan to perform your own Gateway cloning, it is recommended to check for successful transformation of the gene of interest into the vectors. This can be achieved via PCR with either general or specific primers; alternatively, plasmid preps may be made with your kit of choice and the resulting purified DNA sent for sequencing.
3. You will need a concentration of at least 2.5×10^3 mL of zoospores for the detached needle infection assays. The infection conditions have been optimized for *Phytophthora pluvialis* and may need optimization for other *Phytophthora* species.

4. Work quickly (within 2 h) and be gentle when handling the zoospore solution—*P. pluvialis* zoospores encyst very easily when jostled.
5. Selecting post-infection time points to collect samples depends on what protein(s) you are investigating, so be sure to collect at relevant stages.
6. If you wish to harvest samples for RNA extraction at a later time point, flash freeze in liquid nitrogen and store at −80 °C.
7. Treat all equipment with RNaseZAP™ or RNase decontamination solutions. Use only sterile, RNAse-free pipette tips and microcentrifuge tubes; always use proper aseptic technique.
8. Most RNA extraction kits have a maximum amount of tissue you can load onto the accompanying columns. Do not load more than 50 mg of pine tissue onto a column at a time—any greater and you risk clogging the spin column.
9. Do not allow plant tissue samples to thaw—keep in liquid nitrogen until ready to macerate.
10. Do not allow the samples to thaw until addition of lysis buffer.
11. Inspect spin columns after each centrifugation step to ensure all liquid has gone through the column. Residual liquid, particularly ethanol, can inhibit RNA elution. If required, repeat the centrifugation steps.
12. When completing the final drying step prior to RNA elution, ensure absolutely no carryover of ethanol.
13. Keep the microcentrifuge tubes containing the eluted RNA on ice to slow the degradation of the RNA.
14. As RNA is unstable at room temperatures, it can be beneficial to work in small batches when performing RNA extraction to minimize the time the eluted RNA is exposed to the environment.
15. mRNA is roughly 1–5% of the total RNA in a plant cell, so depending on the efficiency of the total RNA extraction, many pine needle/plant tissue replicates ($n = > 60$) may be required if using older needles. Younger pine plantlets will give a higher total RNA yield than mature pine needles.
16. Depending on the brand of mRNA kit, you may need to scale reagents up/down as required; it may be beneficial to concentrate your total RNA beforehand.
17. Allow 5–6 days to create the cDNA library. You may stop at the ethanol precipitation steps after the addition of the glycogen, NH_4OAc and 100% ethanol. In this case store the tubes at −20 °C overnight and resume the following day.
18. Provided with the CloneMiner™ II cDNA Library Construction Kit [18].

19. Sequences for the cDNA construction primers:

    Biotin-attB2- Oligo(dT) Primer Sequence:

    5′-Biotin– GGCGGCCGCACAACTTTGTACAA GAAAGTTGGGT(T)1 9–3′;

    attB1 Adapter Sequences:

    5′-TCGTCGGGGACAACTTTGTACAAAAAAGTTGG-3′

    3′-CCCCTGTTGAAACATGTTTTTTCAACCp-5′

20. Perform size fractionation by column chromatography, as provided with the CloneMiner II kit (Thermo Fisher Scientific). Check that the flow rate is 30–40 s per drop, and the drop size 25–35 μL.
21. A typical yield for the eluted cDNA from the column is around 75–100 ng. If it is lower than this, you can generate a second cDNA library with the saved contents of Tube 3. To check the quantity of the eluted cDNA, we recommend using 1.5 μL of the cDNA on a NanoDrop.
22. Thermo Fisher Scientific recommend using pDONR222 for cDNA construction. In the authors' experience, pDONR221 was more effective than pDONR222.
23. Ensure BP and LR clonases are kept at −20 °C and used within 6 months of receipt. Minimize freeze/thaw cycles to maintain enzyme integrity.
24. When transforming *E. coli* cells, maintain strict sterile technique.
25. There may be fewer clones created than what the CloneMiner II kit suggests. This does not necessarily mean a failed cDNA library creation. Repeat the restriction digest step of the protocol to check for successful transformation.
26. Thermo Fisher Scientific estimate at least 95% recombinants should be seen after the restriction digest. This is the best-case scenario, so you may see a lower number of recombinants (~80%) and still have an efficient cDNA libraryCDNA library [18].
27. If you choose to use the pDONR222 vector, the restriction digest pattern may be slightly different as the *BsrG* I restriction sites are slightly different (442, 1232, 2689) compared to pDONR221 (652, 1442, 2899).

Acknowledgements

This work was supported by grants from Bioprotection Aotearoa and the Radiata Pine Breeding Company.

References

1. Fields S, Song OK (1989) A novel genetic system to detect protein-protein interactions. Nature 340(6230):245–246. https://doi.org/10.1038/340245a0
2. Donnard E, Queiroz EM, Ortega JM, Gietz RD (2014) Yeast two-hybrid liquid screening. Methods Mol Biol 1163:97–107. https://doi.org/10.1007/978-1-4939-0799-1_7
3. Makuch L (2014) Yeast two-hybrid screen. Methods Enzymol 539:31–51. https://doi.org/10.1016/B978-0-12-420120-0.00003-7
4. Chen Q, Wei T (2022) Membrane and nuclear yeast two-hybrid systems. Methods Mol Biol 2400:93–104. https://doi.org/10.1007/978-1-0716-1835-6_10
5. Mitsuda N et al (2010) Efficient yeast one–/two-hybrid screening using a library composed only of transcription factors in *Arabidopsis thaliana*. Plant Cell Physiol 51(12):2145–2151. https://doi.org/10.1093/pcp/pcq161
6. Pelgrom AJE et al (2020) Host interactors of effector proteins of the lettuce downy mildew obtained by yeast two-hybrid screening. PLoS One 15(5):e0226540. https://doi.org/10.1371/journal.pone.0226540
7. Dharmaraj K, Cui W, Rikkerink EHA, Templeton MD (2019) Construction of a kiwifruit yeast two-hybrid cDNA library to identify host targets of the *Pseudomonas syringae* pv. *actinidiae* effector AvrPto5. BMC Res Notes 12:63. https://doi.org/10.1186/s13104-019-4102-x
8. Colombo M et al (2020) NoPv1: a synthetic antimicrobial peptide aptamer targeting the causal agents of grapevine downy mildew and potato late blight. Sci Rep 10:17574. https://doi.org/10.1038/s41598-020-73027-x
9. Du Y et al (2021) RXLR effector PITG20303 targets a potato MKK1 protein to suppress plant immunity. New Phytol 230(2):878–878. https://doi.org/10.1111/nph.17188
10. He Q et al (2019) Effector SFI3 targets potato UBK to suppress early immune transcriptional responses. New Phytol 222(1):438–454. https://doi.org/10.1111/nph.15635
11. Cowell IG (1997) Yeast two-hybrid library screening. Methods Mol Biol 69:185–202. https://doi.org/10.1385/0-89603-383-x:185
12. Horstman A, Tonaco IA, Boutilier K, Immink RG (2014) A cautionary note on the use of split-YFP/BiFC in plant protein-protein interaction studies. Int J Mol Sci 15(6):9628–9643. https://doi.org/10.3390/ijms15069628
13. Asseck LY, Grefen C (2018) Detecting interactions of membrane proteins: the split-ubiquitin system. Methods Mol Biol 1794:49–60. https://doi.org/10.1007/978-1-4939-7871-7_4
14. Meisrimler CN et al (2019) Multiple downy mildew effectors target the stress-related NAC transcription factor LsNAC069 in lettuce. Plant J 99(6):1098–1115. https://doi.org/10.1111/tpj.14383
15. Huang XS et al (2015) ICE1 of *Poncirus trifoliata* functions in cold tolerance by modulating polyamine levels through interacting with arginine decarboxylase. J Exp Bot 66(11):3259–3274. https://doi.org/10.1093/jxb/erv138
16. McLellan H, Armstrong MR, Birch PRJ (2021) Yeast two-hybrid screening for identification of protein-protein interactions in *Solanum tuberosum*. Methods Mol Biol 2354:95–110. https://doi.org/10.1007/978-1-0716-1609-3_4
17. Invitrogen Gateway® Technology (2003) A universal technology to clone DNA sequences for functional analysis and expression in multiple systems. User manual
18. Invitrogen CloneMiner™ II cDNA Library Construction Kit (2012) High-quality cDNA libraries without the use of restriction enzyme cloning techniques. User manual

Chapter 13

Immunoisolation of Plant Endosomal Vesicles to Explore Uptake of Pathogen Effector Proteins During Infection of *Nicotiana benthamiana*

Haixia Wang, Petra C. Boevink, and Paul R. J. Birch

Abstract

Endocytosis is an essential cellular process that uptakes substances into cells at the plasma membrane from the extracellular space and plays a major role in plant development and responses to environmental stimuli. Research has shown that plant membrane-resident proteins are endocytosed and transported into plant endosomes in response to pathogen-secreted elicitors. However, there is no conclusive experimental evidence demonstrating how secreted cytoplasmic effectors from oomycetes and fungi enter host cells during infection. The adapted protocol in this chapter describes endosome isolation using immunopurification with the aim to co-capture *Phytophthora infestans* RXLR effectors during infection. This protocol can be widely used in the isolation and purification of different subtypes of endosomal vesicles that uptake extracellular molecules during pathogen infection and in response to environmental stimuli.

Key words Endocytosis, Oomycetes, Fungi, *Phytophthora*, Late blight, RXLR effector, Immunopurification

1 Introduction

Plants require endocytosis for cell-to-cell communications that mediate hormone and stress signal transduction, uptake of extracellular molecules, and immune responses [1–4]. Endocytosis commences with the packaging of external and plasma membrane proteins/molecules into vesicles, called endosomes, followed by scission from the plasma membrane (PM). The proteins are then predominantly transported via recycling pathways back to the plasma membrane or destined for degradation pathways [5, 6]. There are two endocytic pathways that have been reported in plants: clathrin-mediated endocytosis (CME) and clathrin-independent endocytosis (CIE) [5, 7, 8]. CME has been found to be responsible for endocytosis of PM-resident receptors including brassinosteroid insensitive 1 (BRI1) in *Arabidopsis thaliana*

Monica L. Gerth and Rosie E. Bradshaw (eds.), *Phytophthora: Methods and Protocols*, Methods in Molecular Biology, vol. 2892,
https://doi.org/10.1007/978-1-0716-4330-3_13,

[9, 10], tomato ethylene inducing xylanase receptor (LeEIX2) [11] and the Cf4 receptor, which detects the fungal effector Avr4 [12]. Plant pattern recognition receptors (PRRs), such as FLAGELLIN SENSING 2 (FLS2), elongation factor Tu receptor (EFR), and PEP RECEPTOR 1/2 (PEPR1/2) [2], are also dependent on CME for internalization into plant cells in response to stimulation by the corresponding pathogen-derived ligands. Recent studies reported that endocytic components gather at the extrahaustorial membrane (EHM) [13], including Rab7 and the conserved Rab5 family member Ara7 during oomycete infections [14], and Ara7 and plant-specific ARA6 during infection by the biotrophic powdery mildew fungus *Golovinomyces orontii* [15]. More recently, rice clathrin component OsCLC-1 was also shown to localize to the biotrophic interfacial complex (BIC) formed at the site of cytoplasmic effector delivery from the fungal pathogen *Magnaporthe grisea* [16]. Cytoplasmic effectors from filamentous fungal and oomycete pathogens, such as the well-studied RXLR class from oomycetes, are delivered into living plant cells to suppress immunity [17]. However, the mechanism of host cell entry of translocated filamentous pathogen effectors remains a mystery.

In recent years, the emergence of molecular cell biology and biochemical approaches to isolate sub-cellular compartments is providing insights into the molecular machinery and functions of endocytosis in plants. Confocal microscopy, in combination with genetic techniques and biochemical inhibitor treatments, provides a systems view of trafficking in plant cells. Density gradient centrifugation is a classical method that has been adopted for the isolation of intracellular organelles. However, this method has limitations due to the similar physicochemical properties and densities of diverse sub-cellular compartments that result in them gathering in the same density fractions. The recent exploitation of immunopurification of different vesicle subtypes using specific cellular marker baits in combination with differential ultracentrifugation and proteomic analysis provides a promising strategy to explore and analyse the cargos and components of distinct vesicles [17–22].

In this chapter, we describe in detail modified methods [23] for the capture of *Phytophthora infestans* RXLR effectors that are associated with plant Ara6-coated endosomes during infection, by immunopurification of NbAra6-GFP. Adopting this protocol allows the exploration of entry and subsequent transportation of extracellular molecules into plant vesicles in response to different environmental stimuli.

2 Materials

2.1 Plant Material and Agro-Infiltration

1. Yeast extract and beef (YEB) medium: 5 g/L beef extract, 1 g/ L yeast extract, 5 g/L peptone, 5 g/L sucrose, 0.5 g/L $MgCl_2$.
2. Infiltration medium: 10 mM 2-(N-morpholino) ethanesulfonic acid, 10 mM MgCl2, 150 mM acetosyringone.
3. *Agrobacterium* suspensions carrying a NbAra6-GFP expression vector. NbAra6-GFP was generated as described [23] (*see* **Note 1**).
4. Four- to six-week-old *Nicotiana benthamiana* leaves. Infiltrated with *Agrobacterium* suspension (Subheading. 2.1, **step 1**) (*see* **Note 1**).

2.2 Vesicle Preparation and Fractionation

All buffers, tubes and rotors need to be pre-chilled and kept at 4 °C.

1. 1000× Protease inhibitor cocktail dimethyl sulfoxide (DMSO)-soluble (PIC-D) (*see* **Note 2**): 500 mM phenylmethylsulfonylfluoride (PMSF), 5 mg/mL pepstatin A and 1 mg/ mL chymostatin. Divide into 200 μL aliquots to avoid freeze-thaw. Store at −80 °C.
2. 1000× Protease inhibitor cocktail water-soluble (PIC-W):

   1 M ε-Aminocaproic acid, 5 mg/mL aprotinin and 1 mg/mL leupeptin.

   Divide into 200 μL aliquots to avoid freeze-thaw. Store at −80 °C.
3. Vesicle extraction buffer (VEB): 150 mM Na-HEPES pH 7.5, 10 mM EDTA, 10 mM EGTA, 17.5% (w/w) sucrose (*see* **Note 6**), 7.5 mM KCl,10 mM dithiothreitol, 1× PIC-W and1× PIC-D.
4. 2× SDS-loading buffer: 200 mM Tris-HCl, pH 6.8, 40% glycerol, 2% SDS, 0.04% Coomassie Blue G-250.
5. Plastic boxes with lids. Black plastic sheets.
6. Brix refractometer (*see* **Note 3**).
7. Mortar and pestle.
8. Nylon filter mesh 100 (*see* **Note 4**).
9. Small funnel.
10. 50 mL centrifuge tubes.
11. Refrigerated benchtop centrifuge.
12. Ultracentrifuge, with a fixed-angle 50.2 Ti Rotor.
13. Centrifuge tubes for 50.2 Ti rotor: polycarbonate thick wall centrifuge tubes (25 × 89 mm).
14. Pasteur pipettes.
15. Compact analytical balance.

2.3 Vesicle Immunoisolation

1. GFP magnetic protein A beads (*see* **Note 5**).
2. Binding control magnetic beads (*see* **Note 6**).
3. Magnetic separation rack.
4. Tube rotator.
5. Cold room.
6. Heating block.

2.4 Western Blotting Substrates

1. SuperSignal™ West Femto Maximum Sensitivity Substrate (Catalog number: 34096).
2. Pierce™ ECL Western Blotting Substrate (Catalog number: 32106).
3. Anti-GFP (11814460001, Sigma) raised in mouse at 1:3300 dilution, anti-RFP (5f8-100, Chromotek) raised in rat at 1: 2000 dilution.
4. Ponceau solution (e.g., Thermo Scientific).
5. 5.12% Bis-Tris PAGE gels.
6. Invitrogen™ Mini Gel Tank and Blot Module Set (e.g., Fisher Scientific).
7. Nitrocellulose membrane (e.g., GE Healthcare Life Sciences).
8. 4% milk in 1× PBS 0.1% Tween (1 × PBS-T).

3 Methods

3.1 Agro-Infiltration

1. Transform the NbAra6-GFP expression vector (see Subheading 2.1, **item 1**) into Agrobacterium (*Agrobacterium tumefaciens*) strain GV3101 pMP90 [24].
2. Grow the transformed Agrobacterium cells in YEB medium containing the appropriate antibiotics at 28 °C overnight.
3. Centrifuge the overnight cultures at 3214× *g* for 10 min at room temperature and resuspend pellets in infiltration medium.
4. Incubate the suspension in the dark for at least 1 h.
5. Infiltrate the suspension of *Agrobacterium* harbouring the NbAra6-GFP construct into leaves of 4- to 6-week old *N. benthamiana* plants by wounding then using a 1-mL syringe to pressure-infiltrate, with the *Agrobacterium* cells at a final absorbance at 600 nm (OD_{600}) of 0.005 to 0.1.

3.2 P. infestans Infection

1. Pipette 10 μL drops of *P. infestans* transformant sporangia suspension at 500,000 per mL (*see* **Note 7**) (Fig. 1) onto NbAra6-GFP-expressing leaves 24 h post-Agro-infiltration (*see* **Note 8**).

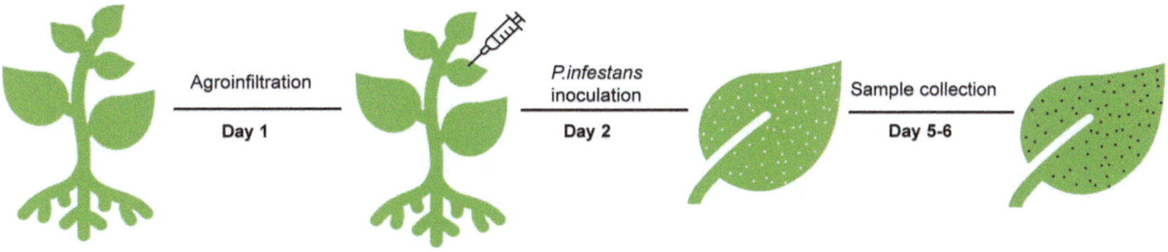

Fig. 1 Schematic representation of plant preparation. Inoculate 10 μL of 500,000 per mL *P. infestans* sporangia drops on *N. benthamiana* leaves 24 h post agroinfiltration. Harvest infected leaves at days 5 and 6 post agroinfiltration

2. Place inoculated leaves in a box with a lid at 19 °C; cover the box with a black sheet for the first 12 h post-inoculation, then remove the black sheet.
3. Harvest infected leaves at 5–6 d post Agro-infiltration. Place leaves into liquid nitrogen and keep samples at −70 °C (*see* **Note 9**) for further analysis.

3.3 Plant Preparation

For immuno-blot analysis, approximately 20 g of infected plant leaves are required. For proteomic analysis, approximately 40 g of infected plant leaves are required (Fig. 1 illustrates plant preparation; *see* **Note 10**).

3.4 Plant Vesicle Preparation

1. Add 2 mL ice-cold VEB to 1 g of plant leaves and grind the leaves gently in a mortar and pestle to a rough slurry until a 5 mL pipette can pipette the slurry smoothly (*see* **Note 11**).
2. Remove the cellular debris by filtration through 100 μm Nylon Filter Mesh and take a 50 μL aliquot as a "Total Input" sample (*see* **Note 12**). Load 10 μL on gel for western blot analysis.

3.5 Vesicle Fractionation by Ultracentrifugation

Plant vesicle extracts are fractionated using ultracentrifugation (Fig. 2 illustrates the fractionation procedure). All steps after harvesting leaves should be performed on ice, and all tubes need to be pre-chilled.

1. Centrifuge the plant vesicle extract at 1000× *g* at 4 °C for 10 min.
2. Transfer supernatants into pre-chilled ultracentrifugation tubes (*see* **Note 13**) and further purify by ultracentrifugation for 40 min at 16,000× *g* using a fixed-angle rotor and polycarbonate thick wall centrifuge tubes.
3. Transfer the supernatants to new ultracentrifuge tubes and centrifuge for 1 h at 30,000× *g* to obtain the P30 pellets. Remove 10 μL for loading on the gel for Western blot.

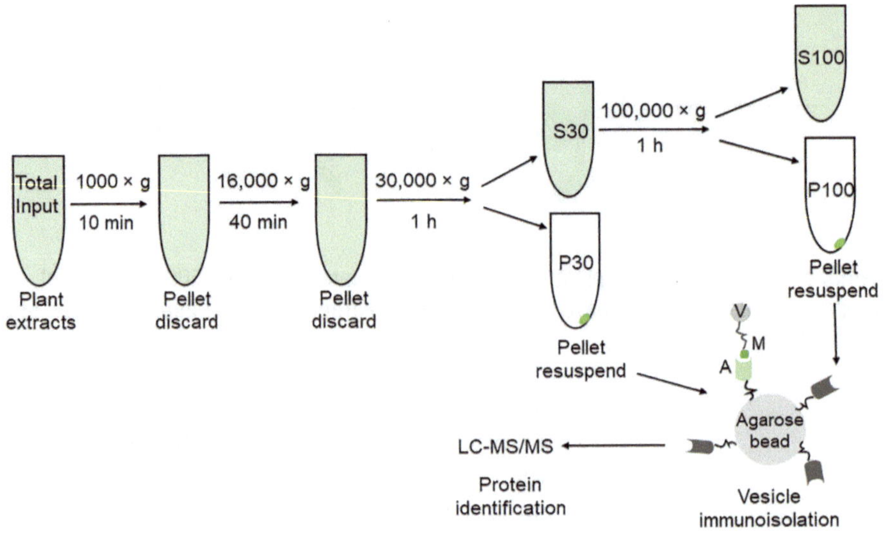

Fig. 2 Schematic representation of vesicle isolation and purification. Plant extracts (Total Input sample) are subjected to differential ultracentrifugation for preparation of crude vesicle fractions (P30 and P100). The enriched crude vesicles are purified using an antibody against the vesicle marker of interest. Following isolation, confocal microscopy and transmission electron microscopy are performed to assess the quality of the enriched vesicles. Following immunoisolation, western blots are performed to test the purity of the purified vesicles and LC-MS/MS analysis is used to identify the proteome. *S* supernatant, *P* pellet, *A* antibody, *M* marker, *V* vesicle

4. Centrifuge the P30 supernatants for 1 h at 100,000× *g* to obtain the P100 pellets. Remove 10 μL for loading on gel for Western blot.
5. Resuspend the pellets in 500 μL VEB buffer by constant rotation for at least 1 h at 4 °C (*see* **Note 14**).
6. Boil 45 μL of resuspended pellets with 45 μL 2 × SDS-loading buffer at 95 °C for 10 min for preparation of the P30 and P100 samples for western blotting.
7. Precipitate proteins in the 800 μL supernatants of P100 with four times the sample volume of pre-chilled acetone overnight at −20 °C, then centrifuge for 10 min at 16,000× *g* and 4 °C as S100 sample. Do not overload S100 sample on gels for Western blot; 10 μL is the maximum.
8. Boil the pellets with 60 μL 2× SDS-loading buffer at 95 °C for 10 min for preparation of the S100 samples for western blotting.

3.6 Immunoisolation of Vesicles

For NbAra6-GFP-labelled vesicle immunoisolation, wash 20 μL GFP magnetic protein A beads and 20 μL binding control magnetic beads with isolation buffer three times for each sample.

1. To pre-clear the suspensions, add 20 μL of the binding control beads to the 1 mL suspensions of P30 or P100, followed by 1 h constant rotation at 4 °C.
2. Transfer these cleared suspensions of P30 or P100 into new tubes and add 20 μL of the washed GFP beads, followed by overnight incubation at 4 °C with constant rotation.
3. Collect the bead-bound endosome material and wash it with isolation buffer at least five times for 2 min at 4 °C with constant rotation (without protease inhibitor cocktails PIC-W or PIC-D).
4. Boil the bead-bound endosomes with 45 μL 2× SDS-loading buffer.
5. Resolve proteins by western blotting. Separate proteins on 12% Bis-Tris PAGE gels using an X-blot Mini Cell (Thermo Fisher Scientific), followed by transfer to nitrocellulose membrane (GE Healthcare Life Sciences) using an X10 Blot Module (Thermo Fisher Scientific) according to the manufacturer's instructions. Stain membranes with Ponceau solution to confirm transfer and even loading. Membranes were blocked in 4% milk in 1× PBS 0.1% Tween (1× PBS-T) with shaking for 1 h at room temperature before incubation with the appropriate antibodies overnight (*see* **Note 15**).

4 Notes

1. This method uses NbAra6-GFP as the marker of endosome immunopurification. The C-terminal GFP fused NbAra6 allows the interaction of the antibody and GFP tag during immunoisolation, and the NbAra6 N-terminus attaches to the endosome membranes and allows NbAra6 positive endosomes to be pulled down with GFP Trap magnetic beads [23]. Special consideration should be taken when selecting the endosome marker that will be used as a "bait" protein for immunoisolation of endosomes; in particular, the interaction with the antibody and the fused tag must not disrupt its function.
2. 1000× PIC-D easily freezes on ice, so keep it at room temperature.
3. A refractometer is required to measure the sucrose concentration.
4. 100 μm Nylon Filter Mesh or Miracloth.
5. Instead of commercial magnetic beads, commercial agarose beads and specific antibody-conjugated beads can be used.

6. Binding control beads can be used as pre-clearing beads and negative controls to check the level of nonspecific background in immunoprecipitated samples.
7. For *Agrobacterium*-infiltrated plants, the given number of *Phytophthora infestans* spores per 10 μL should be enough to get early infection at 5–6 d post-Agro-infiltration. After this time, infected leaves are not ideal for the experiment. For transgenic and wild-type plants and for other pathogen species, one would need to test infection in advance to determine the final inoculum concentration. It is very important to leave about 5 mm space between spore suspension drops. Be aware that macroscopically visible lesions are not ideal areas to capture effectors as they contain large numbers of necrotic cells. Colonized areas with invisible lesions are the best for the capture of effectors. Do not have the spore suspension fully covering leaves or spray spore suspension onto leaves.
8. Test protein expression levels in advance to determine the lowest final infiltration concentration (OD_{600}) of *Agrobacterium* to avoid having protein aggregations and other overexpression artefacts during infection. Instead of agroinfiltrated plants with the tagged endosome markers of interest, transgenic plants that carry tagged markers can be used. Alternatively, if specific antibodies for endosome makers are available, wild type plants are the ideal plant materials.
9. Freshly infected leaves are the ideal materials for endosome isolation. Freezing leaves with liquid nitrogen and keeping at −80 °C allows the collection of enough infected plant material for endosome isolation by combining multiple infection assays.
10. It is important to have enough starting plant material. This protocol aims to capture the uptake of effector proteins into plant cells; only a small portion of them will be captured during infection. The amount of plant tissue may vary depending on the downstream analyses.
11. It is very important to keep the mortar, pestle, tubes, and VEB cold. If leaves are frozen, take leaves from the freezer and transfer to a pre-cooled mortar immediately and grind the leaves into small pieces to evaporate liquid nitrogen and then add VEB to avoid having frozen VEB and leaves. For efficient vesicle isolation, the maximum weight of leaves for each sample grind is 5 g.
12. If quantification and comparison of results are required, the "Total Input" can be precipitated with acetone.
13. It is recommended to use 10 mL lysate for each ultracentrifugation tube and to fill the tube with VEB to allow large compartments to easily sediment while minimizing the loss of microvesicles caused by vesicle aggregation.

14. Well-resuspended samples are very important for the efficiency of bead binding.

15. To test the signal of C-terminal RFP-tag fused effector, load 10 μL Input sample and 20 μL IP sample on an SDS-PAGE gel. Leave an empty well between Input and IP samples to avoid contamination from Input samples during development of ECL film. In our study, we used a mixture of SuperSignal™ West Femto Maximum Sensitivity Substrate: Pierce™ ECL Western Blotting Substrate in a ratio = 1:2 for developing films as the IP effector signal is quite weak. We used Pierce™ ECL Western Blotting Substrate for GFP and other antibodies. Anti-RFP: dilution 1:2000, Anti-Rat: dilution 1:5000. In our previous studies, Ara6-GFP was detected in all samples and P30 gave the strongest signal (23). Effector-RFP was barely detectable in the Total Input sample, and undetectable in S100 sample. P30 Input and P100 Input samples showed clear bands for effector-RFP protein. P30 IP showed a visible effector-RFP band, P100 IP samples gave a very weak band for effector-RFP (23). Effector-RFP signal strength depends on the efficiency of secretion of the effector and the level of infection.

Acknowledgements

This work was financially supported by the Biotechnology and Biological Sciences Research Council grant BB/S003096/1; European Research Council (ERC)-Advanced grant PathEVome (787764); and Scottish Government Rural and Environment Science and Analytical Services Division (RESAS) JHI B1-1.

References

1. Ekanayake G, LaMontagne ED, Heese A (2019) Never walk alone: Clathrin-coated vesicle (CCV) components in plant immunity. Annu Rev Phytopathol 57:387–409
2. Mbengue M et al (2016) Clathrin-dependent endocytosis is required for immunity mediated by pattern recognition receptor kinases. Proc Natl Acad Sci USA 113(39):11034–11039
3. Fan L, Li R, Pan J, Ding Z, Lin J (2015) Endocytosis and its regulation in plants. Trends Plant Sci 20(6):388–397
4. Chen X, Irani NG, Friml J (2011) Clathrin-mediated endocytosis: the gateway into plant cells. Curr Opin Plant Biol 14(6):674–682
5. Paez Valencia J, Goodman K, Otegui MS (2016) Endocytosis and endosomal trafficking in plants. Annu Rev Plant Biol 67:309–335
6. Reynolds GD, Wang C, Pan J, Bednarek SY (2018) Inroads into internalization: five years of endocytic exploration. Plant Physiol 176:208–218
7. McMahon HT, Boucrot E (2011) Molecular mechanism and physiological functions of clathrin-mediated endocytosis. Nat Rev Mol Cell Biol 12(8):517–533
8. Sandvig K, Kavaliauskiene S, Skotland T (2018) Clathrin-independent endocytosis: an increasing degree of complexity. Histochem Cell Biol 150:107–118
9. Di Rubbo S et al (2013) The clathrin adaptor complex AP-2 mediates endocytosis of brassinosteroid insensitive1 in *Arabidopsis*. Plant Cell 25(8):2986–2997
10. Liu D et al (2020) Endocytosis of brassinosteroid insensitive1 is partly driven by a canonical

Tyr-based motif. Plant Cell 32(11): 3598–3612

11. Sharfman M, Bar M, Ehrlich M, Schuster S, Melech-Bonfil S, Ezer R, Sessa G, Avni A (2011) Endosomal signaling of the tomato leucine-rich repeat receptor-like protein LeEix2. Plant J 68(3):413–423
12. Postma J, Liebrand TW, Bi G, Evrard A, Bye RR, Mbengue M, Kuhn H, Joosten MH, Robatzek S (2016) Avr4 promotes Cf-4 receptor-like protein association with the BAK1/SERK3 receptor-like kinase to initiate receptor endocytosis and plant immunity. New Phytol 210(2):627–642
13. Bozkurt TO, Kamoun S (2020) The plant-pathogen haustorial interface at a glance. J Cell Sci 133(5):jcs237958
14. Lu YJ, Schornack S, Spallek T, Geldner N, Chory J, Schellmann S, Schumacher K, Kamoun S, Robatzek S (2012) Patterns of plant subcellular responses to successful oomycete infections reveal differences in host cell reprogramming and endocytic trafficking. Cell Microbiol 14:682–697
15. Inada N et al (2016) Modulation of plant RAB GTPase-mediated membrane trafficking pathway at the interface between plants and obligate biotrophic pathogens. Plant Cell Physiol 57(9):1854–1864
16. Oliveira-Garcia E, Tamang TM, Park J, Dalby M, Martin-Urdiroz M, Rodriguez Herrero C, Vu AH, Park S, Talbot NJ, Valent B (2021) Clathrin-mediated endocytosis facilitates internalization of *Magnaporthe oryzae* effectors into rice cells. Plant Cell 35(7): 2527–2551
17. He Q, McLellan H, Boevink PC, Birch PRJ (2020) All roads lead to susceptibility: the many modes of action of fungal and oomycete intracellular effectors. Plant Commun 1(4): 100050
18. Steuble M, Gerrits B, Ludwig A, Mateos JM, Diep TM, Tagaya M, Stephan A, Schätzle P, Kunz B, Streit P, Sonderegger P (2010) Molecular characterization of a trafficking organelle: dissecting the axonal paths of calsyntenin-1 transport vesicles. Proteomics 10(21): 3775–3788
19. Jeppesen DK, Fenix AM, Franklin JL, Higginbotham JN, Zhang Q, Zimmerman LJ, Liebler DC, Ping J, Liu Q, Evans R, Fissell WH, Patton JG, Rome LH, Burnette DT, Coffey RJ (2019) Reassessment of exosome composition. Cell 177(2):428–445
20. He B, Cai Q, Qiao L, Huang CY, Wang S, Miao W, Ha T, Wang Y, Jin H (2021) RNA-binding proteins contribute to small RNA loading in plant extracellular vesicles. Nat Plants 7(3):342–352
21. Suzuki SW, Emr SD (2022) Immunoisolation of endosomal recycling vesicles from *Saccharomyces cerevisiae.* Bio-protocol 12(9):e4403
22. Suzuki SW, Oishi A, Nikulin N, Jorgensen JR, Baile MG, Emr SD (2021) A PX-BAR protein Mvp1/SNX8 and a dynamin-like GTPase Vps1 drive endosomal recycling. elife 10:e69883
23. Wang H, Wang S, Wang W, Xu L, Welsh LRJ, Gierlinski M, Whisson SC, Hemsley PA, Boevink PC, Birch PRJ (2023) Uptake of oomycete RXLR effectors into host cells by clathrin-mediated endocytosis. Plant Cell 35(7): 2504–2526
24. Koncz C, Martini N, Mayerhofer R, Koncz-Kalman Z, Körber H, Redei GP, Schell J (1989) High-frequency T-DNA-mediated gene tagging in plants. Proc Natl Acad Sci USA 86(21):8467–8471

Chapter 14

Fabrication of Lab-on-a-Chip Devices to Study the Forces Imparted by Growing *Phytophthora* Hyphae

Ayelen Tayagui, Yiling Sun, Volker Nock, and Ashley Garrill

Abstract

In this chapter, we describe the design and manufacture of a Lab-on-a-Chip (LoC) device suitable for measuring the μN forces exerted by tips of growing *Phytophthora* hyphae. LoC describes microfluidic devices, typically made of the polymer polydimethylsiloxane (PDMS), that are increasingly being used to answer fundamental questions in biological, chemical, physical, and medical research. These LoC devices enable the integration of several laboratory functions on small plastic devices that are quick to produce and easy to replicate. While we describe the design and manufacture of one particular device, the methods covered are applicable to any chip design, and such devices could be used to investigate many aspects of *Phytophthora* biology.

Key words Lab-on-a-Chip, Microfluidic devices, PDMS, *Phytophthora*, Force measurement

1 Introduction

Lab-on-a-Chip (LOC) is a term that is used to describe microfluidic devices that have the capability of integrating one or more laboratory functions [1, 2]. The devices allow for high-throughput analyses in a convenient format, typically a microscope slide, and are quick and easy to produce by accurate and cheap replication from a master mold. While their initial design and fabrication can be technically demanding, their many advantages, such as the ability to compartmentalize single cells and ease of sample observation, mean that they have found widespread usage in biology, chemistry, physics, and medical research, and in the clinic, where they offer cheap and rapid point-of-care testing [1, 2]. For oomycetes and fungi, their use has included the measurement of mechanical forces generated in yeast and hyphal cells [3–6], the maze-solving and dynamic behavior of basidiomycete hyphae [7], the thigmotropic responses and wall stiffness of *Candida albicans* [8, 9],

Monica L. Gerth and Rosie E. Bradshaw (eds.), *Phytophthora: Methods and Protocols*, Methods in Molecular Biology, vol. 2892, https://doi.org/10.1007/978-1-0716-4330-3_14,

high-throughput screening using yeast cell cultures [10, 11], and the detection of *Phytophthora* species [12].

There is no one set design for the devices, as they can be made to address specific questions. For plant pathogens such as *Phytophthora,* it is possible to design devices to study different aspects of the invasive life cycle [13]. In general, a device might include an experimental chamber or channels in which the growth/behavior of hyphae or spores can be observed; feeding into these would be input/output channels linked to seeding/outlet ports that allow the initial inoculation of *Phytophthora* onto the devices and the addition/removal of media/experimental treatments. In our laboratory, we have incorporated elastomeric pillars into growth channels that have been used to measure μN forces as oomycete and fungal hyphae grow into them [4–6], electrodes that generate electric fields to study zoospore electrotaxis [14], pneumatic microvalves to trap individual zoospores and measure the forces exerted by emerging germ tubes [15, 16], and gas channels that enable the generation of oxygen gradients to study aerotropic growth and responses to hypoxia [17].

Microfluidic devices are commonly made from the silicone elastomer polydimethylsiloxane (PDMS). PDMS is biocompatible, permeable to gas, and has excellent mechanical and optical properties allowing for flexibility and high-quality imaging of cells. Initially, devices are designed using computer-aided design (CAD) software (Fig. 1a–d), and these designs are then transferred to chrome-on-glass photomasks using a pattern generator (Fig. 2a, b). This is the initial step in a manufacturing process called photolithography, as part of which a photosensitive polymer (photoresist) is selectively exposed to light through a mask, leaving a latent image in the polymer that can then be selectively dissolved to provide a molding structure for replica-casting [18, 19]. In the example shown in Fig. 2, 18 replicates of a device are reproduced on a single photomask. Relatively simple devices incorporating constant-height structures can be created using just one photomask, whereas more complex ones incorporating stepped features require additional photomasks (Fig. 2a, b). The photomasks are used to create molds (Figs. 2c and 3). This process requires UV light-sensitive photoresists that are coated to a desired thickness onto a silicon wafer (Fig. 3). The photoresist is then positioned underneath the photomask using a mask aligner and exposed to UV light and cured. There are two types of photoresists, referred to respectively as positive-tone and negative-tone. Positive-tone photoresists are removed where they have been exposed to UV light once they are developed, whereas negative-tone photoresists remain. Thus, it is possible to reproduce either positive or negative replicas of the photomask design into a photoresist. These replicas then serve as the mold onto which the PDMS prepolymer is poured to produce the device.

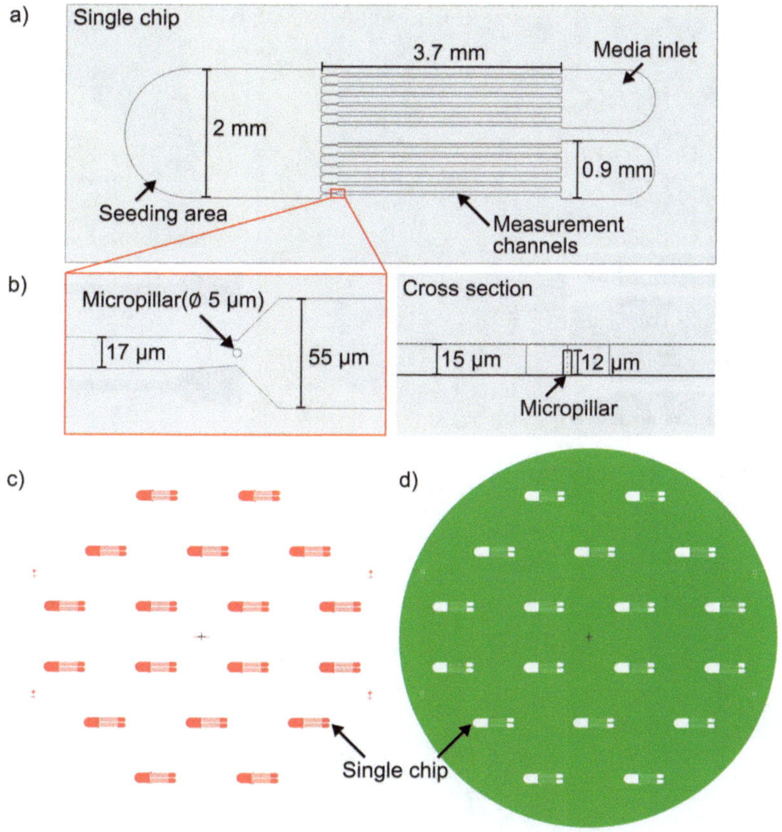

Fig. 1 Schematics illustrating a typical chip design in CAD software. A top-down view of a single chip (**a**) is used to define the layout and dimensions of measurement channels, seeding area, and media inlets. While measurement channels include a circle defining the diameter of the measurement micropillar (**b**), the heights of channels and pillars, as shown in cross-section, are not defined in the design file but rather by choosing a certain thickness of photoresist during mold fabrication. Single chips are multiplied and arrayed for export into two mask files by designing two independent layers in CAD (**c, d**). These differ only in the absence (**c**) or presence (**d**) of the pillar circle, and their image polarity, which is determined by the type of photoresist used (**c** negative-tone, **d** positive-tone). Note: Dimensions shown are aimed at hyphae with widths ranging from 8 to 15 μm, and are given as reference values for our own LoC designs

In the following methodology, we describe the fabrication of a device that enables the measurement of forces exerted by individual *Phytophthora* hyphae. The device consists of a seeding area into which hyphae can be inoculated using an agar mycelial plug. Emanating from this are 12 parallel channels, which comprise an initial tapered area leading to a narrower portion that contains a single elastomeric pillar. Beyond the pillar, the channel widens and leads to a media supply area that allows the addition of media (Fig. 2d).

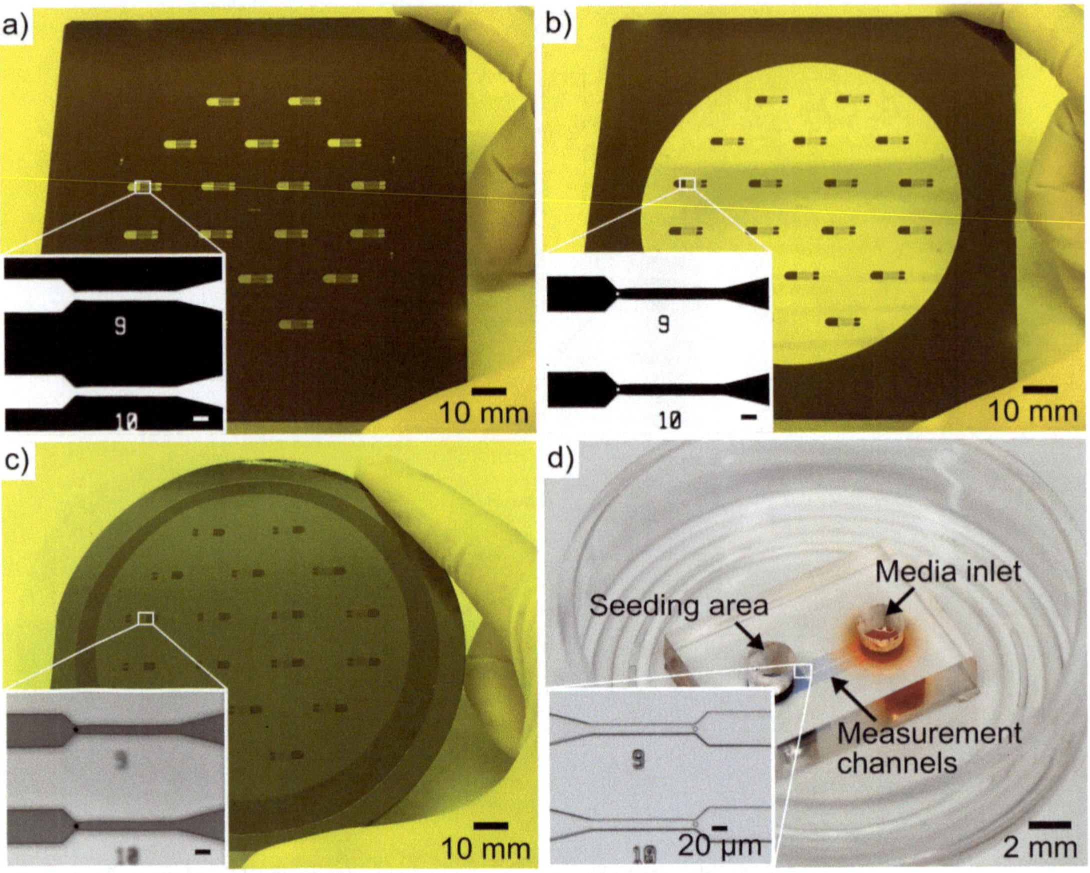

Fig. 2 Images of the two 4″ chrome-on-glass photomasks (**a, b**), the silicon wafer with photoresists to which the designs from the photomasks have been transferred (**c**) and the final device bonded to a fluorodish (**d**). The first negative layer contains the channel outlines and seeding/media inlet areas, with the inset showing two tapered growth channel designs (**a**). The second positive layer contains the channel outlines, seeding/media inlet areas and measurement pillars. The inset shows the outline of two channels and pillars (**b**). The silicon wafer with the transferred patterns from the photomasks. The inset shows two channels and pillars (**c**). A completed device bonded to a fluorodish with the seeding/media inlets and channels indicated. These have been filled with colored food dye to provide better contrast. The inset shows two channels and elastomeric pillars (**d**)

This device requires two photomasks; the first negative-tone layer contains the seeding and media supply areas and channels (Fig. 2a) and the second positive-tone layer, the seeding and media supply areas, channels, and elastomeric pillars (Fig. 2b). Two layers are needed so that the top of the replicated PDMS pillar can move freely when a force is applied to it. Prior to use, the device is prefilled with growth media and inoculated with hyphae, which grow from the seeding area along the channels. The tapered area directs the hyphae toward the pillar and the narrow channel around the pillar increases the chance that the hyphae first contact the pillar with their tip. By optically tracking the displacement of the pillar

[20], this device enables the estimation of the force exerted by hyphal tips [4–6, 15, 16]. With relatively straightforward modifications to the design of the device, other aspects of the invasive life cycle of *Phytophthora* species should be amenable to study.

2 Materials

2.1 Preparation of the Masks

1. A computer-aided design (CAD) program (e.g., L-Edit, Siemens Software).
2. Laser mask writer (e.g., uPG101, Heidelberg Instruments).
3. Chrome-on-glass photomask blank (e.g., 4″, Nanofilm).
4. Tetramethylammonium hydroxide (TMAH) based developer AZ326MIF (Microchemicals GmbH).
5. Deionized water.
6. Compressed nitrogen.
7. Chromium etchant solution.
8. Methanol.
9. Acetone.
10. Isopropyl alcohol.

2.2 Mold Fabrication

1. Negative-tone photoresist AZ15nXT (Microchemicals GmbH).
2. Positive-tone photoresist AZ12XT (Microchemicals GmbH).
3. 4″ silicon wafer.
4. Spin coater (e.g., WS-650, Laurell Technologies).
5. Contact hotplate (e.g., HP30, Torrey Pines Scientific).
6. Mask aligner (e.g., MA6, Suss MicroTec).
7. Mask aligner filter (e.g., PL-360, Omega Filters).
8. Tetramethylammonium hydroxide (TMAH) based developer AZ326MIF (Microchemicals GmbH).
9. Deionized water.
10. Compressed nitrogen.

2.3 Replica Casting

1. Trichloro (1H, 1H, 2H, 2H-perfluorooctyl) silane (e.g., #448931, Sigma-Aldrich).
2. Polydimethylsiloxane (PDMS) solution. To prepare, mix the base and curing agent of a silicone-based elastomeric kit (e.g., Sylgard 184 Silicone Elastomer Kit, Dow Corning) at a ratio of 10:1 (w/w). Alternative silicones: Sylgard-186 (Dow Corning), MED 10-6400 (NuSil), ExSil 100 (Gelest), and Elastosil-M4630 and M4130 (Wacker Chemicals).

3. Biopsy punch (e.g., T983-30, Rapid-Core, ProSciTech).
4. Vacuum chamber (e.g., Vacuum Desiccator PP/PC, Cole Parmer).
5. Vacuum pump (e.g., Rocker 600, Rocker).
6. Contact hotplate (e.g., HP30, Torrey Pines Scientific).

2.4 Bonding of Devices to Transparent Substrates

1. Glass substrates (e.g., Fluorodishes (FD35-100, World Precision Instruments), glass coverslips, or standard microscope slides) (*see* **Note 1**).
2. Vacuum chamber (e.g., Vacuum Desiccator PP/PC, Cole Parmer).
3. Vacuum pump (e.g., Rocker 600, Rocker).
4. Barrel asher (e.g., Emitech K1050X, Quarum Technologies or Tergeo, PIE Scientific).
5. Contact hotplate (e.g., HP30, Torrey Pines Scientific).
6. Vacuum sealer (e.g., Sunbeam food saver).

2.5 Inoculation of Devices and Force Measurement

1. *Phytophthora* culture growing as mycelia on agar (e.g., V8 agar).
2. Clarified V8 broth: Dissolve 1.5 g $CaCO_3$ into 150 mL V8 juice by stirring for at least 15 min. Clarify the juice by centrifuging at 1073× *g* for 15 min. To 100 mL of the clarified supernatant, add 900 mL deionized water. Add 0.05 g of β-sitosterol and mix. Autoclave to sterilize. Store at 4 °C.
3. Microscope with a camera (*see* **Note 2**).

3 Methods

3.1 Design of the Device and Preparation of the Masks

Use a computer-aided design (CAD) program (L-Edit, Siemens Software) to design the layout of the device (Fig. 1, *see* **Note 3**). Klayout, AutoCAD (Autodesk), or any other CAD tool are possible software alternatives, but they typically require more attention when the output is to be converted into the mask writer file format.

1. Transfer the design to a laser mask writer for transfer of the designs to chrome-on-glass photomasks.
2. Load a chrome-on-glass photomask blank into the laser mask writer.
3. Expose the design onto the blank.
4. Develop the photomask in TMAH-based developer AZ326MIF for 1 min.
5. Rinse the photomask with deionized water to stop the development, then dry with a nitrogen gun.

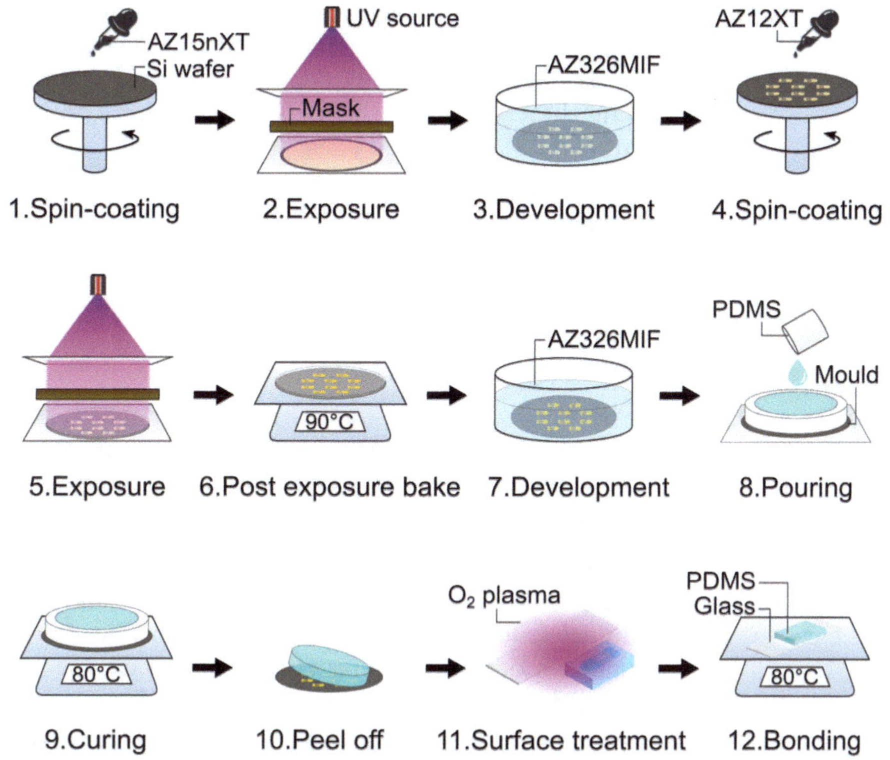

Fig. 3 The various steps in the production of a PDMS device using soft lithography. Photoresist is spread to a desired thickness on a silicon wafer using a spin coater (**step 1**). The wafer is positioned next to the photomask from Fig. 1a in a mask aligner and exposed to UV light (**step 2**). The wafer is then baked, developed, and washed (**step** 3). For more complex devices, a second photoresist and photomask are used to introduce additional attributes of the device design. Photoresist is spread to a desired thickness on top of the silicon wafer using a spin coater and then positioned next to the second photomask in a mask aligner (**steps** 4 and 5). The wafer is then baked, developed, and washed (**steps** 6 and 7). PDMS is poured onto the wafer (**step** 8), cured (**step** 9), and then carefully peeled off (**step** 10). The PDMS devices thus produced are bonded to glass slides or fluorodishes using oxygen plasma and baking (**steps** 11 and 12)

6. In a fume hood, submerge the fully developed photomask in chromium etchant solution, with manual agitation, for 1 min.
7. Rinse the photomask with deionized water to stop etching, then dry with a nitrogen gun.
8. To remove any remaining photoresist, immerse the photomask in acetone in a glass container and sonicate for 5 min, followed by the same procedure for methanol and then isopropanol.
9. Dry with a nitrogen gun.

3.2 Mold Fabrication

1. Spin-coat a first layer of negative-tone photoresist AZ15nXT that is 2.6 μm thick onto a clean 4″ silicon wafer using a spin coater set at 5000 rpm (Fig. 3, **step 1**).
2. Soft-bake the silicon wafer for 2 min at 110 °C on a hotplate.

3. Align the first layer mask (Fig. 2a) and photoresist-laminated silicon wafer in a mask aligner set to vacuum contact mode with a mask aligner filter present, and expose to UV-light (365 nm) using the UV lamp of the mask aligner with a dose of 90 mJ/cm^2 (Fig. 3, **step 2**).
4. Post-exposure, bake the wafer for 1 min at 120 °C on a contact hotplate.
5. Develop the wafer in TMAH-based developer AZ326MIF for 50 s, rinse with deionized water, and dry using a nitrogen gun (Fig. 3, **step 3**).
6. Hard bake the wafer for 10 min at 200 °C on a contact hotplate.
7. Spin coat a second layer of positive-tone photoresist AZ12XT 12 μm thick on top of the first layer on the wafer using the spin coater set at 1500 rpm (Fig. 3, **step 4**).
8. Then, soft bake for 3 min at 110 °C on a hotplate.
9. Align the second layer mask (Fig. 2b) and photoresist-laminated silicon wafer in the mask aligner set in the vacuum contact mode with a filter (PL-360) and expose to UV-light with a dose of 260 mJ/cm^2 (Fig. 3, **step 5**).
10. Post-exposure, bake the wafer for 1 min at 90 °C on a contact hotplate (Fig. 3, **step 6**).
11. Develop the wafer, which now contains the transferred patterns from both the first and second layer photomasks, by immersion into a TMAH-based developer (e.g., AZ326MIF) for 2 min. Then, rinse the wafer with deionized water and dry using nitrogen (Fig. 3, **step 7**).

3.3 PDMS Replica-Casting

1. Before PDMS casting, treat the mold by exposure to trichloro (1H, 1H, 2H, 2H-perfluorooctyl) silane vapour in a vacuum for 2 h to facilitate mold release.
2. Mix the PDMS base with curing agent at a 10:1 (w/w) ratio, degas, and then pour onto the mold (Fig. 3, **step 8**).
3. Bake for 2 h at 80 °C on a hotplate, then peel off the cured PDMS from the mold (*see* **Note 4**) and bake for an additional 4 h at 80 °C to ensure the desired hardness is reached (Fig. 3, **steps 9** and **10**).
4. Cut out the individual devices using a guillotine or a scalpel.
5. Create holes in the device with a 3 mm diameter biopsy punch to provide access to the seeding and media supply areas.

3.4 Bonding of Devices to Fluorodishes/ Microscope Slides

1. Depending on the microscope that will be used to observe the cells, bond the device to the bottom of a fluorodish or microscope slide using a 30 s exposure to 100 W oxygen plasma in a barrel asher (Fig. 3, **step 11**).
2. Strengthen the bonding to the dish or slide by baking for 2 h at 80 °C on a hotplate (Fig. 3, **step 12**).
3. Place the devices in a vacuum chamber and degas for 2 h to prepare for vacuum-assisted filling [21].
4. Optional: Degassed devices can be sealed into food-grade vacuum bags using a vacuum sealer and stored until use.

3.5 Inoculation of Hyphae on the Device and Force Measurements

1. Remove the device from the vacuum-sealed bag and immediately fill inlets with V8 broth. Examine through the microscope to make sure there are no air bubbles throughout the device (*see* **Note 5**).
2. Take a 3 mm mycelial plug from the growing edge of a *Phytophthora* culture using a biopsy punch.
3. Using a scalpel, transfer the agar plug (with the hyphae on the bottom of the plug) into the seeding area of the device. Position the inoculum such that the hyphal tips face the entrance of the channels.

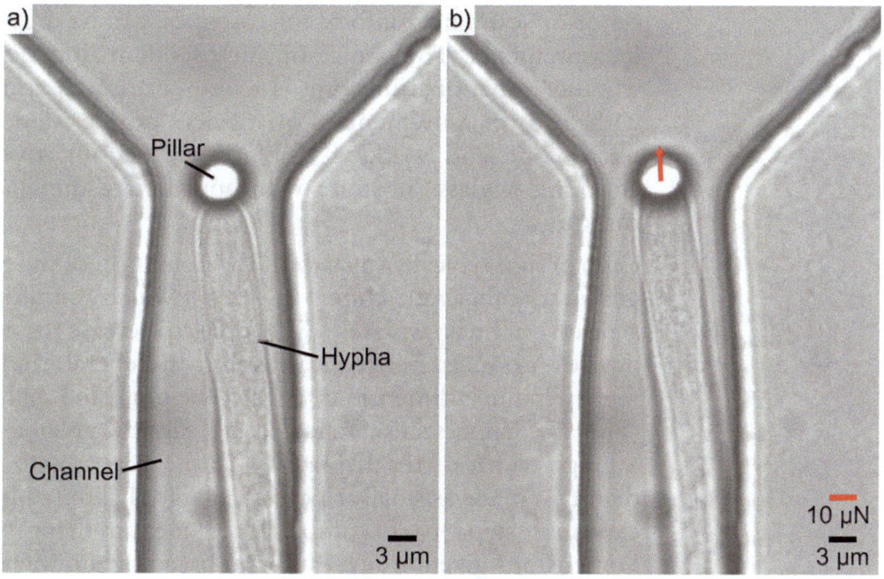

Fig. 4 A hypha of *Phytophthora palmivora* growing down a growth channel on a LoC device towards an elastomeric pillar. (**a**) The narrowing of the channel around the pillar increases the chances of the hyphal tip hitting the pillar. Optical tracking of the displacement of the pillar enables an estimation of the force imparted by the hypha as it hits the pillar. (**b**) The red arrow indicates the magnitude and direction of the force

4. Once inoculated, keep the device in the dark at room temperature for the duration of an experiment, typically between 8 and 12 h, depending on the growth rate of the hyphae (*see* **Note 6**).
5. Monitor growth every 30 min and, if necessary, top up the media supply inlet with V8 broth to avoid evaporation (*see* **Note 6**).
6. When the tips of the hyphae are around 100 μm from the pillars, begin recording so that the deflection of the top of the pillars can be measured over time (Fig. 4a, b).
7. Convert the pillar deflections into force measurements (Fig. 4b, *see* **Note** 7).

4 Notes

1. The choice of substrate depends on the microscope and imaging technique used, as well as how many measurement channels are to be included in one LoC device. For example, fluorodishes enable imaging on inverted microscopes using higher magnification objective lenses with a shorter working distance but provide only a limited bonding area and, thus, a reduced overall chip size. Microscope slides provide a larger bonding area but may be too thick for objective lenses with a short working distance.
2. Inverted and upright light microscopes, as well as epifluorescence or scanning confocal microscopes, can be used for force measurements. The choice of microscope mainly depends on whether functional imaging is to be performed concurrently with force measurements. LoC devices can be adapted to the microscope and working distance of the objective lens by choosing a glass substrate with appropriate dimensions and thickness.
3. Chip design typically starts with the definition of the top-down geometry of a single chip (Fig. 1a, b), which is multiplied and arrayed on the final masks (Fig. 1c, d) to increase the number of devices produced per replica casting. In general, the measurement channel dimensions (width/height) and pillar height (Fig. 1b, cross-section) should be slightly greater than the average width of the hyphae to be measured, with the pillar diameter made as small as possible (typically limited by the UV exposure step) to maximize force measurement sensitivity. Dimensions given in Fig. 1a, b are based on a device designed to measure hyphae with widths ranging from 8 to 15 μm [6], and can be used as starting values for other designs. The height of the pillar and the vertical gap between its top and the glass

substrate are defined by the thickness of the two resist layers used and their respective masks (Fig. 1c, d). It is important to include the circle defining the pillar only on the second mask (Fig. 1d), as this guarantees a vertical separation of the pillar top from the glass substrate by not creating a pillar hole in the first layer (Fig. 1c), enabling it to freely move when actuated by a hyphal tip. This is typically achieved by designing two independent layers in CAD (Fig. 1c, d), which differ only in the absence (Fig. 1c) or presence (Fig. 1d) of the pillar circle, and exporting these as separate mask files.

4. To prevent the breaking of the pillars, care is needed when peeling the PDMS off the wafer mold. Pillars should release from the mold by themselves when the PDMS is slowly angled upwards. For this, the PDMS should be at room temperature.
5. Once devices are removed from the vacuum-sealed bags prior to use, they should be filled immediately with liquid growth media to better prevent the formation of air bubbles in the channels and chambers. If air bubbles are present, these can usually be expelled by perfusing more media through the device.
6. For longer experiments, ensure that the growth media inlet is constantly refilled and does not dry out. It helps if the device is kept in the dark, except at times when observations/images are being made/taken. This ensures that the microscope light does not heat up the growth media, affect the growth rate of hyphae, and/or increase evaporation.
7. For force conversion, the x/y-coordinates of the pillar top need to be tracked over time. This can be done manually by recording the x- and y-pixel coordinates of the pillar top centre over the recorded image sequence or semi-automatically using tools such as the ImageJ plugin TrackMate [22]. These coordinates are then converted to micrometers according to the scale of the recorded images. Finally, the recorded pillar top coordinates combined with the device parameters, such as material properties and geometry, are used to convert movement into force values using a mechanical model for a bending pillar [4]. This conversion can be implemented as a simple script in various software environments such as Matlab [23] or Python [24]. An example implementation in the form of a Jupyter Notebook [25] is available via GitHub [26].

References

1. Azizipour N, Avazpour R, Rosenzweig DH, Sawan M, Ajji A (2020) Evolution of biochip technology: a review from lab-on-a-chip to organ-on-a-chip. Micromachines 11:599. https://doi.org/10.3390/mi11060599
2. Convery N, Gadegaard N (2019) 30 years of microfluidics. Micro Nano Eng 2:76–91. https://doi.org/10.1016/j.mne.2019.01.003

3. Minc N, Boudaoud A, Chang F (2009) Mechanical forces of fission yeast growth. Curr Biol 19:1096–1101. https://doi.org/10.1016/j.cub.2009.05.031
4. Tayagui A, Sun Y, Collings DA, Garrill A, Nock V (2017) An elastomeric micropillar platform for the study of protrusive forces in hyphal invasion. Lab Chip 17(21):3643–3653. https://doi.org/10.1039/C7LC00725F
5. Sun Y, Tayagui A, Shearer H, Garrill A, Nock V (2018) A microfluidic platform with integrated sensing pillars for protrusive force measurements in *Neurospora crassa*. In: IEEE 31st international conference on micro electro mechanical systems (MEMS), Belfast, Northern Ireland, p 1116–1119. https://doi.org/10.1109/MEMSYS.2018.8346756
6. Sun Y, Tayagui A, Garrill A, Nock V (2018) Fabrication of in-channel high-aspect ratio sensing pillars for protrusive force measurements on fungi and oomycetes. J Microelectromech Syst 27:827–835. https://doi.org/10.1109/JMEMS.2018.2862863
7. Hanson KL, Nicolau DV Jr, Filipponi L, Wang L, Lee AP, Nicolau DV (2006) Fungi use efficient algorithms for the exploration of microfluidic networks. Small 2:1212–1220. https://doi.org/10.1002/smll.200600105
8. Thomson DD, Wehmeier S, Byfield FJ, Janmey PA, Caballero-Lima D, Crossley A, Brand AC (2015) Contact-induced apical asymmetry drives the thigmotropic responses of *Candida albicans* hyphae. Cell Microbiol 17:342–354. https://doi.org/10.1111/cmi.12369
9. Couttenier E, Bachellier-Bassi S, d'Enfert C, Villard C (2022) Bending stiffness of *Candida albicans* hyphae as a proxy of cell wall properties. Lab Chip 22(20):3898–3909. https://doi.org/10.1039/D2LC00219A
10. Sjostrom SL, Bai Y, Huang M, Liu Z, Nielsen J, Joensson HN, Svahn HA (2014) High-throughput screening for industrial enzyme production hosts by droplet microfluidics. Lab Chip 14(4):806–813. https://doi.org/10.1039/C3LC51202A
11. Kou S, Cheng D, Sun F, Hsing IM (2016) Microfluidics and microbial engineering. Lab Chip 16(3):432–446. https://doi.org/10.1039/C5LC01039J
12. König S, Schwenkbier L, Pollok S, Riedel M, Wagner S, Popp J, Weber K, Werres S (2015) Potential of Ypt1 and ITS gene regions for the detection of *Phytophthora* species in a lab-on-a-chip DNA hybridization array. Plant Pathol 64: 1176–1189. https://doi.org/10.1111/ppa.12357
13. Sun Y, Tayagui A, Sale S, Sarkar D, Nock V, Garrill A (2021) Platforms for high-throughput screening and force measurements on fungi and oomycetes. Micromachines 12: 639. https://doi.org/10.3390/mi12060639
14. Sarkar D, Sun Y, Tayagui A, Adams R, Garrill A, Nock V (2022) Microfluidic platform to study electric field based root targeting by pathogenic zoospores. In: IEEE 35th international conference on micro electro mechanical systems conference (MEMS), Tokyo, Japan, p 884–887. https://doi.org/10.1109/MEMS51670.2022.9699559
15. Sun, Y, Tayagui, A, Garrill, A, Nock, V (2019) A monolithic polydimethylsiloxane platform for zoospore capture, germination and single hypha force sensing. In: IEEE 20th international conference on solid-state sensors, actuators and microsystems & eurosensors XXXIII (TRANSDUCERS & EUROSENSORS XXXIII), Berlin, Germany, p 409–412. https://doi.org/10.1109/TRANSDUCERS.2019.8808505
16. Sun Y, Tayagui A, Garrill A, Nock V (2020) Microfluidic platform for integrated compartmentalization of single zoospores, germination and measurement of protrusive force generated by germ tubes. Lab Chip 20(22):4141–4151. https://doi.org/10.1039/D0LC00752H
17. Tayagui A, Sun Y, Garrill A, Nock V (2023) A microfluidic oxygen gradient generator for the study of aerotropism in hyphae of Oomycetes. In: IEEE 36th international conference on micro electro mechanical systems (MEMS), Munich, Germany, p 997–1000. https://doi.org/10.1109/MEMS49605.2023.10052442
18. Xia YN, Whitesides GM (1998) Soft lithography. Angew Chem Int Ed 37(5):551–575
19. Xia YN, Kim E, Zhao X-M, Rogers JA, Prentiss M, Whitesides GM (1996) Complex optical surfaces formed by replica molding against elastomeric masters. Science 273(5273):347–349. https://doi.org/10.1126/science.273.5273.347
20. Ghanbari A, Nock V, Johari S, Blaikie RJ, Chen X, Wang W (2012) A micropillar-based on-chip system for continuous force measurement of *C elegans*. J Micromech Microeng 22(9):095009. https://doi.org/10.1088/0960-1317/22/9/095009
21. Hosokawa K, Sato K, Ichikawa N, Maeda M (2004) Power-free poly(dimethylsiloxane) microfluidic devices for gold nanoparticle-based DNA analysis. Lab Chip 4(3):181–185. https://doi.org/10.1039/B403930K
22. Tinevez J-Y, Perry N, Schindelin J, Hoopes GM, Reynolds GD, Laplantine E et al (2017)

TrackMate: an open and extensible platform for single-particle tracking. Methods 115:80–90. https://doi.org/10.1016/j.ymeth.2016.09.016

23. The MathWorks, Inc (2023) MATLAB version: 9.14.0 (R2023a)
24. Rossum GV, Drake FL (2009) Python 3 reference manual: createspace
25. Granger BE, Pérez F (2021) Jupyter: thinking and storytelling with code and data. Comput Sci Eng 23(2):7–14. https://doi.org/10.1109/MCSE.2021.3059263
26. Nock V (2023) Pillar force calculation (v1.0.0). Zenodo. https://doi.org/10.5281/zenodo.8404864

Chapter 15

Application of Proteomic Methods in Oomycete Biology

Christina E. Andronis, Scott Bringans, and Kar-Chun Tan

Abstract

The biochemical makeup of any organism provides insight into key factors regarding its biological functions. These factors can be explored using proteomics, which allows us to obtain a snapshot of the protein content and abundance in an organism, cell type or sub-cellular compartment. Here, we describe proteomic methodologies that can be used to dissect the biochemical mechanism of phytopathogenicity in oomycetes. These methodologies include protein extraction, purification, subsequent processing, mass spectrometry analysis, and qualitative and quantitative data processing of oomycete proteomes for comparative studies. Additionally, the use of mass spectra to assist in gene validation and modelling in unfinished oomycete genomes is also described.

Key words Proteomics, Proteogenomics, Proteome, Mass spectrometry, Oomycete

1 Introduction

The biochemical makeup of any organism provides insight into key factors regarding their roles in survival and relationships with other organisms. These factors can be explored using multi-omic approaches that reflect their genomic, transcriptomic, metabolomic, and proteomic composition. Tools and technologies designed to elucidate these components enable a better understanding of how and why organisms behave and respond in certain ways. This drives the progression of formulating new innovative management strategies that apply across many fields, from human disease to crop production.

Proteomics is a powerful tool that enables the protein content (or proteome) of an organism to be obtained in a particular state, which can help inform on function [1]. The proteome of an organism is the collection of proteins that are found in an organism in a particular state concerning localization, treatments, and interactions at a particular time [1]. Proteome mapping using mass spectrometry is the most commonly used method for identifying proteins in a biological sample. The continual improvement of

Monica L. Gerth and Rosie E. Bradshaw (eds.), *Phytophthora: Methods and Protocols*, Methods in Molecular Biology, vol. 2892, https://doi.org/10.1007/978-1-0716-4330-3_15,

Table 1
Advantages and limitations of label-free and labelled quantitative proteomics. Isobaric Tag for Absolute and Relative Quantification (iTRAQ). Tandem Mass Tags (TMT). Stable Isotope Labelling by Amino acids in Cell Culture (SILAC)

	Advantages	Limitations
Label-free quantification	No specialized preparation Lower total protein amount required Fractionation is not possible (retention-time dependence) Instrument time < 3 h	Lower reproducibility Lack of peptide normalisation
Labelled quantification (iTRAQ, TMT, SILAC)	Lower variability of quantification Multiplexing Reduced technical variability Fractionation is used to separate peptides more thoroughly leading to greater proteome coverage	A higher protein amount required Instrument time > 24 h due to multiple fractions to be analyzed Costly labels

analytical technology results in greater sensitivity to the point where proteomics now has a greater level of accessibility in the molecular plant pathology community.

A major advantage of proteomics over other quantitative tools, such as transcriptomics, is that it can be used to identify, quantify, and monitor changes across different tissue/cell types, treatments, and sub-cellular compartments with the aim of gaining an insight into biochemical processes occurring at the time of sampling [2–4]. These proteomic pipelines rely on high-quality protein extraction, where the proteins are subsequently proteolytically cleaved into peptides. For quantitative proteomics, peptides can be analyzed as label-free or labelled with an isobaric tag (Table 1).

Methods of separation can be implemented at this stage to gain depth of proteome coverage. Separated peptides are subsequently analyzed by mass spectrometry to obtain mass spectra of fragmented peptides. The mass spectra are then matched to an annotated database of the organism and, depending on the experimental design, can be used for protein identification, mapping, and quantification. Dissecting the proteome by enrichment of sub-proteomes of cell types and cellular compartments allows for a greater diversity of protein detection and validation of a higher proportion of the gene annotation [5–7]. Quantitative proteomics can be used to study the differential abundance of proteins, creating a more detailed assessment of these complex systems [8, 9].

Proteomics is highly reliant on genomic data for identification, where mass spectra are matched to the genome annotation. The importance of high-quality genome annotations to facilitate accurate characterization of protein content and function should be noted. Proteomics has traditionally been used for profiling and comparative analysis of protein abundance. However, alternative use of spectral data has brought about proteomic-assisted genome annotation. Discrepancies of correlation between transcripts and proteins have provided a unique opportunity to use mass spectra to manually curate genes. Termed "proteogenomics," this approach provides a complementary approach to other gene calling methods such as in silico gene prediction and RNAseq validation [10].

Agricultural industries and natural environments are facing constant threats from a suite of phytopathogenic oomycetes. Destructive phytopathogenic oomycetes include those from the genera *Pythium, Phytophthora, Peronospora, Bremia, Albugo, Aphanomyces, Sclerospora, Lagena,* and *Plasmopara* [11–13]. The affected plants span from crops and horticulture to native vegetation and forests globally. They contribute to significant losses from reduced crop yield, associated resource wastage (including land, water, and fertilizer), pesticide application, and destruction of natural ecosystems [14].

Proteomics has been used to unravel the cell-specific proteomes of several oomycete species, including the *Phytophthora* species *infestans, palmivora, sojae, capsica,* and *cinnamomi* [2, 15–17]. These include comparative biochemical analyses of hyphal/infection structures and elucidation of the extracellular proteomes where most host-pathogen interaction molecules such as effectors are found [7, 15–20].

In this chapter, we describe proteomic pipelines as applied to oomycetes and explain how to obtain biologically relevant information from spectral data. In this chapter, we begin by describing a protein extraction protocol for oomycete samples, followed by processing from protein to mass spectrometry-compatible peptides. Options for labelled proteomics have also been included. We define a liquid chromatography-mass spectrometry (LC-MS/MS) pipeline for the acquisition of mass spectra and subsequent data processing to yield qualitative and quantitative outputs as well as peptide data that can be exploited for gene model refinement through proteogenomics [3, 4, 17, 21, 22]. Although the protocol described below was optimized for *Phytophthora cinnamomi*, it is applicable to other oomycetes and filamentous fungal pathogens [17, 23–25].

2 Materials

2.1 Starting Material

1. Oomycete samples can include isolated sub-cellular compartments or cell types such as hyphae, zoospores, and the secretome that can be obtained from axenic growth conditions (Fig. 1a). Several published protocols can be used to isolate or enrich for particular oomycete cell types in vitro [4, 17, 26]. Up to 10 mg of dry-weight mycelia and $4e^5$ zoospores will

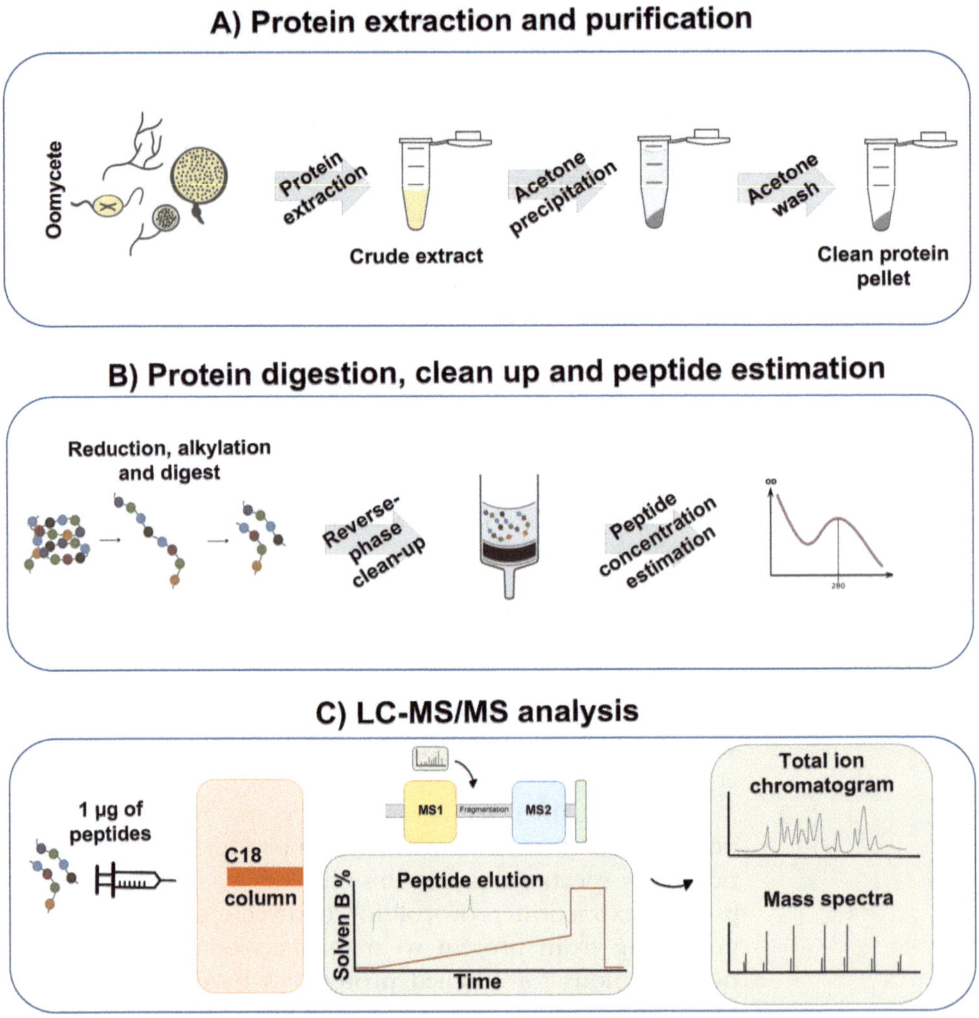

Fig. 1 A workflow for protein identification, quantification, and proteogenomic output from oomycete and/or host plant samples. (**a**) The proteins of oomycete tissue samples are extracted and further purified by acetone precipitation. (**b**) The purified protein pellet is subsequently reduced, alkylated then digested using trypsin. This peptide mixture is desalted, and the concentration of the resulting purified peptide mixture is estimated by reading the optical density at 250 nm. (**c**) 1 µg of peptides is loaded onto the analytical column and peptides are eluted over a 190-min gradient and analyzed by mass spectrometry

yield approximately 500 µg of proteins. Mycelial tissue can be obtained from V8 agar plates, but the protein extraction methods can be used on oomycete tissue grown on other solid or liquid complexes, or defined media. Approximately 200 mL of culture filtrate yields approximately 250 µg of secreted proteins [17]. Do not grow the mycelia in liquid culture for more than 3 days (or stationary growth phase) for the production of the secretome as a prolonged period results in starvation and causes excessive intracellular contamination into the secretome due to cell death and lysis. Filter with gauze or by passing the secretome through a 0.2 μm syringe filter to remove all debris from the culture media.

2. Mixed samples such as oomycete-plant tissue (e.g., infected root and leaves) can also be used [27].
3. All samples should be snap-frozen and freeze-dried as soon as they have been harvested to minimize protein degradation.
4. We recommend that all sample groups be prepared in biological triplicates as a minimum number of replicates. Mycelia from three plates should suffice for each replicate. Secretomes from four 50 mL flasks of culture filtrate should suffice for each replicate.

2.2 Protein Extraction

1. Liquid nitrogen.
2. Plastic pestles and safe-lock 1.5 mL microcentrifuge tubes OR mortar and pestle, OR metal beads and bead shaker (e.g., Retch, Haan, Germany).
3. Oomycete protein extraction buffer: 25 mM Tris–HCl pH 7.5, 0.25% (w/v) SDS, 50 mM sodium phosphate, 1 mM sodium fluoride, 50 μM sodium orthovanadate, 1 mM phenylmethylsulfonyl fluoride, and a protease inhibitor cocktail (e.g., P8215 Sigma, St Louis USA).
4. Mixed infection protein extraction buffer: 125 mM Tris–HCl, pH 7.0, 7% SDS, 0.5% (w/v) polyvinylpyrrolidone 40 (PVP-40).
5. Ice-cold methanol.
6. Ice-cold chloroform.
7. Ice-cold acetone.
8. Freeze drier.

2.3 Protein Digestion and Clean-up

1. 0.5 M triethylammonium bicarbonate (TEAB).
2. Urea buffer: 6 M urea in 0.5 M TEAB.
3. 50 mM tris ((2-carboxyethyl)phosphine) (TCEP).
4. 200 mM methyl methanethiosulfonate (MMTS).
5. 1 μg/μL sequencing grade trypsin (e.g., from bovine pancreas, Sigma).
6. Strata-X 33 μm polymeric reverse phase column (e.g., Phenomenex).
7. LC-MS grade methanol.
8. LC-MS grade acetonitrile.
9. Elution buffer: 80% acetonitrile (LC-MS grade), 0.1% formic acid.

10. Vacuum concentrator (Eppendorf SpeediVac or equivalent).
11. Nanodrop spectrophotometer.

2.4 Peptide Labelling (Optional) with iTRAQ or TMT

1. iTRAQ (Isobaric Tags for Relative and Absolute Quantification) 4-plex or 8-plex depending on sample numbers (Sciex).
2. 0.5 M TEAB, pH 7 (pH with HCl).
3. TMT (Tandem Mass Tags) 6–18- plex (Thermo Fisher Scientific).
4. 100 mM TEAB pH 8 (pH with HCl).
5. 5% Hydroxylamine/ 100 mM TEAB.

2.5 LC-MS/MS

1. Loading Buffer A: 2% LC-MS grade acetonitrile, 0.1% formic acid in LC-MS grade water.
2. Loading Buffer B: 98% LC-MS grade acetonitrile, 0.1% formic acid in LC-MS grade water.
3. Thermo Ultimate 3000 RS liquid chromatography system (Thermo Scientific) configured to nano-mode, or equivalent HPLC.
4. Acclaim™ PepMap™ 100 C18 LC column, 2 μm particle size × 150 nm (Thermo Scientific) or equivalent column.
5. Acclaim™ PepMap™ 100 C18 trapping column 300 nm (Thermo Scientific) equivalent.
6. Q-Exactive HF mass spectrometer (Thermo Scientific) or equivalent fusion or Q-TOF MS/MS system (*see* **Note 1**).

2.6 Software and Bioinformatic Tools

1. Proteome Discoverer software (Thermo Scientific, version 2.4.1.15 or later) or similar software such as MaxQuant (Max Plank Institute of Biochemistry), Protein Pilot (Sciex), Mascot (Matrix Science), Trans-Proteomic Pipeline (Institute for Systems Biology), or Proteo Wizard (Proteo Wizard Software Foundation) (**Note 2** and Table 2) [28].
2. Appropriate protein and gene/CDS databases for the organism of interest (assembly scaffold fasta, protein fasta, gene/ CDS gff3).
3. Geneious Prime (https://www.geneious.com/) or UGENE (http://ugene.net/) or equivalent.
4. Python (https://www.python.org/).
5. EMBOSS getorf (version 6.6 or later) (https://www.bioinformatics.nl/cgi-bin/emboss/getorf).
6. BEDtools (version 2.28.0 or later) (https://bedtools.readthedocs.io/) CDS Mapper (version 0.6, 2011, https://sourceforge.net/projects/cdsmapper/).
7. BLASTp (version 2.9.0 or later) and tBLASTn (version 2.11 or later) (https://blast.ncbi.nlm.nih.gov/Blast.cgi).

Table 2
Example analysis software for application in shotgun qualitative and quantitative proteomics

Software package	Accessibility	Dataset types	Specialized features
Protein Pilot (Sciex)	Commercial licence	Qualitative, iTRAQ, TMT (up to 6-plex)	Provides high quality spectral matching with comprehensive false discovery reporting [36].
Proteome Discoverer (Thermo Fisher)	Commercial licence	Qualitative, label-free, iTRAQ, TMT, SILAC	Allows for integration of study design including replicates and variables into the workflow [28].
Max Quant (Max Plank Institute)	Freely available	Qualitative, label-free, iTRAQ, TMT, SILAC	Provides a higher specificity, accuracy and precision compared to other packages [37].
Mascot (Matrix Science)	Freely available	Qualitative, label-free, iTRAQ, TMT, SILAC	Utilizes a robust scoring algorithm for protein identification [38].
Trans-proteomic Pipeline (Institute for Systems Biology)	Freely available	Qualitative, iTRAQ, SILAC	Operates on a complement of tools providing alternatives for fit-for-purpose analysis [39].

3 Methods

3.1 Protein Extraction

Proteomics can be used to take a biochemical snapshot of a particular sample type, such as mycelia, zoospores, secreted proteins, or mixed samples [3, 4, 17, 21, 22]. Methods for protein extraction are described for these different types of samples in Subheading 3.1.1, 3.1.2, and 3.1.3, with some general considerations for all extractions in Subheading 3.1.4. We routinely use a Nanodrop reading at 280 nm to generate an estimated protein and peptide concentration of the sample after extraction and tryptic digest, respectively. However, BCA and Bradford methods can also be used [29]. For mycelial, zoospore and secretome protein extraction using methods outlined in this chapter, estimation at the peptide level is used. This provides a more accurate estimation for relative quantification; however, this method of estimation can also be used at the protein level if required.

3.1.1 Mycelia and Zoospore Protein Extraction

1. Freeze-dry the isolated and snap-frozen oomycete sample.
2. Using a plastic pestle and 1.5 mL microcentrifuge tube (or alternatives as listed in **step 2** of Subheading 2.3), grind the freeze-dried sample into a fine powder. Samples should be kept cold during this process.
3. Add 200–300 μL of oomycete protein extraction buffer. This should form a light slurry.

4. Incubate on ice for 30 min with regular gentle mixing.
5. Centrifuge at 20,000× *g* at 4 °C for 30 min. Decant the supernatant containing the protein fraction into a new 1.5 mL tube. Ensure no cell debris is decanted, as this will affect downstream processing.
6. Proceed to acetone precipitation (see Subheading 3.1.5).

3.1.2 Secretome Protein Extraction

1. For the secretome, add 50 μL of protease inhibitor cocktail for every 50 mL of culture filtrate before snap-freezing and freeze-drying.
2. Reconstitute the freeze-dried culture filtrate in 3 mL of water and mix gently by vortexing.
3. Proceed to acetone precipitation (see Subheading 3.1.5).

3.1.3 Mixed Infection Sample Protein Extraction (I.E., Infected Root and Leaves)

1. Freeze-dry the snap-frozen infected plant tissue samples.
2. Using a plastic pestle and 1.5 mL microcentrifuge tube (or alternatives as listed in **step 2** of Subheading 2.3), grind the freeze-dried sample into a fine powder. Samples should be kept cold during this process.
3. Add 300 μL of mixed infection protein extraction buffer. This should form a light slurry.
4. Incubate on ice for 30 min with regular gentle mixing.
5. Centrifuge at 20,000× *g* at 4 °C for 30 min. Decant 200 μL of the supernatant containing the protein fraction into a new 1.5 mL tube. Ensure no cell debris is decanted, as this will affect downstream processing.
6. Add 800 μL of ice-cold methanol and 200 μL of ice-cold chloroform. Vortex and add 500 μL of water. Centrifuge samples for 5 min at 15,000× *g* at 4 °C.
7. Discard the upper aqueous phase, careful not to disrupt the protein pellet.
8. Add an additional 500 μL of ice-cold methanol. Invert the tube and carefully discard the supernatant.
9. Proceed to acetone precipitation (see Subheading 3.1.5).

3.1.4 General Considerations for Protein Extraction from All Tissue Types

Protein denaturation can occur during this process if the material is not handled on ice, or the process is not performed rapidly. To ensure protein integrity is maintained, samples can be tested using one-dimensional (1D) sodium dodecyl sulphate-polyacrylamide gel electrophoresis (Fig. 2) [23, 30]. This is a common method in biochemistry to determine the degradation of protein by observing the banding patterns of proteins. If the gel appears to be smeared and down the whole lane, then the proteins are degraded (Fig. 2d). If well-defined bands are observed, then the proteins are intact (Fig. 2a–c).

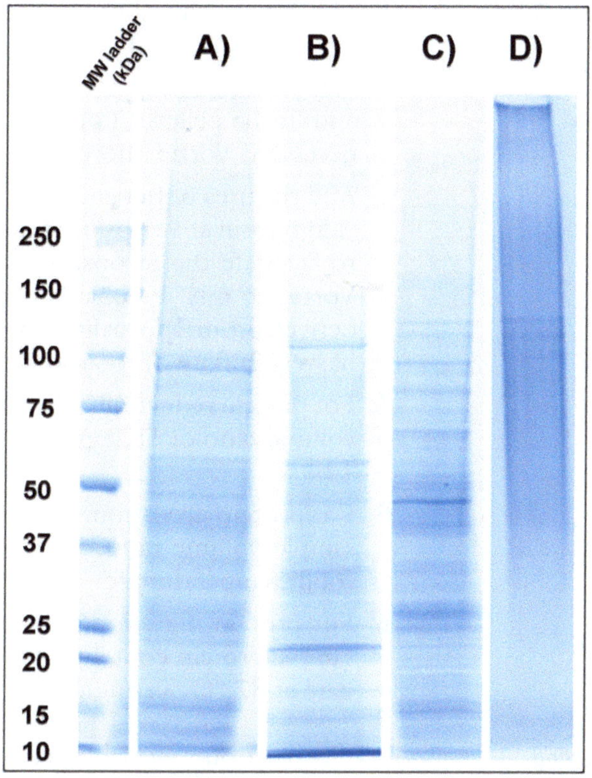

Fig. 2 SDS-1D gel example. Lanes a, b, and c show examples of good-quality sub-proteome extracts. The differing gel band patterns indicate differing protein content. Lane d is an example of a poor-quality extraction, as indicated by the smearing down the column and ill-defined bands

3.1.5 Acetone Precipitation of Proteins

1. To the crude extract of oomycete protein (or to the protein pellet for mixed infection samples), add six volumes of ice-cold acetone.
2. Invert the tube to mix and incubate at −20 °C for 16 h.
3. Pulse spin and discard the acetone, being careful not to disturb the protein pellet. For the secretome, carefully transfer the pellet to a 1.5 mL tube (as the secretome material was harvested and precipitated in a 50 mL centrifuge tube) using a small spatula or pipette tip. Ensure any residual pellet from the centrifuge tube or spatula/ pipette tip is rinsed off into the 1.5 mL tube with acetone.
4. Wash the protein pellet by adding 1 mL of acetone, invert the tube to mix, pulse spin, and discard the acetone. Do not let the pellet dry out completely, as this will make it difficult to re-solubilize the protein. This leaves a pellet of purified protein (Fig. 1a).

3.2 Protein Pellet Digestion

1. The entire pellet should be used for subsequent processing.
2. To the protein pellet, add 0.5 M TEAB. The solution volume may vary depending on the pellet size. Start by adding 200 μL and, if the pellet is not submerged, add additional solution if necessary. Vortex thoroughly to mix.
3. Add the urea buffer at a buffer-to-sample volume ratio of 1:10. Add trypsin at a trypsin-to-sample protein content ratio of 1:40 to facilitate the reconstitution of the pellet into the solution. Vortex to mix. Incubate at 37 °C for 3 h. For protein pellets derived from approximately 250 mg of tissue, we routinely harvest 200 μg of total soluble proteins.
4. For protein reduction, add 50 mM TCEP at a TCEP-to-sample volume ratio of 1:10. Vortex to mix and incubate at 37 °C for 1 h.
5. For protein alkylation, add 200 mM MMTS at an MMTS-to-sample volume ratio of 1:20. Vortex to mix and incubate at room temperature for 10 min.
6. For protein digestion, add additional trypsin at a trypsin-to-sample protein content ratio of 1:10. Vortex to mix and incubate at 37 °C overnight (Fig. 1b).

3.3 Desalting Samples

Full instructions available at: https://www.phenomenex.com/ViewDocument?id=strata-xl+general+method+for+the+extraction+of+neutral+compounds+(tn-0029)&fsr=1

Desalt samples using Strata-X 33 μm polymeric reverse-phase columns (Phenomenex). Allow the liquid to drain in between the addition of each reagent.

1. Place a waste collection tube below the column.
2. Add 2 × 1 mL of methanol.
3. Add 2 × 1 mL of acetonitrile.
4. Add 2 × 1 mL of water.
5. Add 0.5 mL of elution buffer.
6. Add 2 × 1 mL of water.
7. Add sample.
8. Add 2 × 1 mL of water.
9. Place a 1.5 mL microcentrifuge tube below the column for sample collection.
10. Add 0.5 mL of elution buffer to elute the sample into the collection tube.
11. Dry the peptides using a vacuum centrifuge.

3.4 Peptide Concentration Estimation

1. Reconstitute the dried peptides in 100 μL of LC-MS/MS loading buffer A.
2. Blank a Nanodrop (or equivalent) with LC-MS/MS loading buffer A, set for "Protein Quantification" and read the sample absorbance at 280 nm (Fig. 1b). Measure the concentration of each sample in triplicate, ensuring the sample pedestal is cleaned between each reading.
3. For unlabelled qualitative and unlabelled proteomics, proceed to Subheading 3.6.
4. For labelled quantification, dry peptides using a vacuum centrifuge and proceed to Subheading 3.5.

3.5 Labelled Quantification

1. For labelled quantification (such as isobaric tags for relative and absolute quantification (iTRAQ) (Sciex) and tandem mass tags (TMT) (Thermo Fisher), label each sample for comparison as per the manufacturer's instructions (*see* **Note 3**) [25]. Peptide concentration estimation is required here to ensure equal total peptide amounts for all samples are labelled. See Subheading 3.6 Peptide concentration estimation. In this case, reconstitute the samples in the required buffer for labelling and use this to blank the nanodrop. Below are the procedures for iTRAQ (Subheading 3.5.1) and TMT (Subheading 3.5.2) as examples of peptide labelling for quantitative proteomics.

3.5.1 iTRAQ Labelling (Sciex)

Full labelling instructions available at: https://sciex.com/products/consumables/itraq-reagent

1. Bring reagents to room temperature and add 70 μL of ethanol to each iTRAQ Reagent vial. Vortex and pulse spin.
2. Make up each sample in 0.5 M TEAB, pH 7 (pH with HCl) to a final sample concentration of 3.33 mg/mL.
3. Take 30 μL of each sample (100 μg) and add it to an iTRAQ label. Be sure to note which sample was added to each label. Vortex, pulse spin, and incubate for 1 h at room temperature.
4. Add 1 mL of water and incubate for 30 min.
5. Combine labelled samples into one 15 mL centrifuge tube and desalt using Strata-X 33 μm polymeric reverse-phase columns (see Subheading 3.2).
6. Dry the labelled peptides using a vacuum centrifuge.
7. Labelled peptides should be separated by reverse phase high pH chromatography and concatenated into 12 fractions (*see* **Note 4**).

3.5.2 TMT Labelling (Thermo Fisher)

Full labelling instructions available at: https://www.thermofisher.com/au/en/home/life-science/protein-biology/protein-mass-spectrometry-analysis/protein-quantitation-mass-spectrometry/tandem-mass-tag-systems.html

1. Bring reagents to room temperature and add 41 μL of acetonitrile. Vortex and pulse spin.
2. Make up each sample in 100 mM TEAB, pH 8 (pH with HCl) to a final sample concentration of 1.0 mg/mL.
3. Take 100 μL of each sample (100 μg) and add it to a TMT label. Be sure to note which sample was added to which label. Vortex, pulse spin, and incubate for 1 h at room temperature.
4. Add 8 μL of 5% hydroxylamine/100 mM TEAB and incubate for 15 min at room temperature.
5. Combine labelled samples into one 15 or 50 mL tube (depending on the number of samples) and desalt using Strata-X 33 μm polymeric reverse-phase columns (*see* Subheading 3.3).
6. Dry the labelled peptides using a vacuum centrifuge.
7. Labelled peptides should be separated by reverse phase high pH chromatography and concatenated into 12 fractions (*see* **Note 4**).

3.6 Mass Spectrometry Analysis of Peptides

1. In this step, 1 μg of the sample is loaded onto the Acclaim™ PepMap™ 100 C18 LC column. Dilute the sample concentration if necessary and adjust the injection volume accordingly.
2. Run samples on the mass spectrometer over a 190-min linear gradient starting from 3% Loading Buffer A to 40% Loading Buffer B. Peptides typically elute between 10% and 35% acetonitrile (Fig. 1c, Tables 3 and 4).

Table 3
HPLC conditions

Injection volume	1 μL (or equivalent for 1 μg on column)	
Run time	190 min	
Autosampler temp.	4 °C	
Flow rate	300 nL/min	
Oven temperature	40 °C	
Gradient profile	Time (min)	Solvent B % (98% ACN/ 0.1% formic acid)
	0	3
	5	3
	165	6
	179	45
	180	98
	190	3

Table 4
Mass spectrometry conditions based on the QE-HE Exactive mass spectrometer (Thermo Scientific)

Ion source	Nanospray flex
Polarity	Positive
Capillary temperature	250 °C
Ion spray voltage	1.50 KV
Scan range	350–1400 m/z
Full MS[1]	
Resolution	120,000 FWHM
AGC target[a]	3e6
Maximum IT[b]	50 ms
Data-dependent-MS[2]	
Resolution	30,000 FWHM
AGC target[a]	2e5
Maximum IT[b]	55 ms

[a]Automatic gain control: desired number of charges in the orbitrap MS analyzer
[b]Limits the time of accumulating ions per scan event

3.7 Generation of Spectral Matches Using Proteome Discoverer

1. The output from high-resolution nano-LC-MS/MS can be analyzed through various data interpretation software. Here we describe the analysis of spectral data using Proteome Discoverer (Thermo Scientific). Other software packages can also be used (Table 2).
2. Create a new study and use the template processing and consensus workflows. Select the appropriate workflow depending on the type of analysis required, as shown in Fig. 3a.
3. Variables such as replicates and categorical factors (such as sample treatments and experimental conditions) should be included in the study definition (Fig. 3a). This is particularly useful for quantitative analysis. Import the raw files into the study and assign each sample with its respective study variables. For labelled quantification, the quantification channels should be set for each sample during this step (i.e., which label was added to which sample).
4. Each node of the processing and consensus workflows has parameters that can be altered to fit the experimental design and desired analysis output. Most of the parameters can be kept at default, but modifications (in the SequestHT node, which is used for spectral matching) such as oxidation, deamidation,

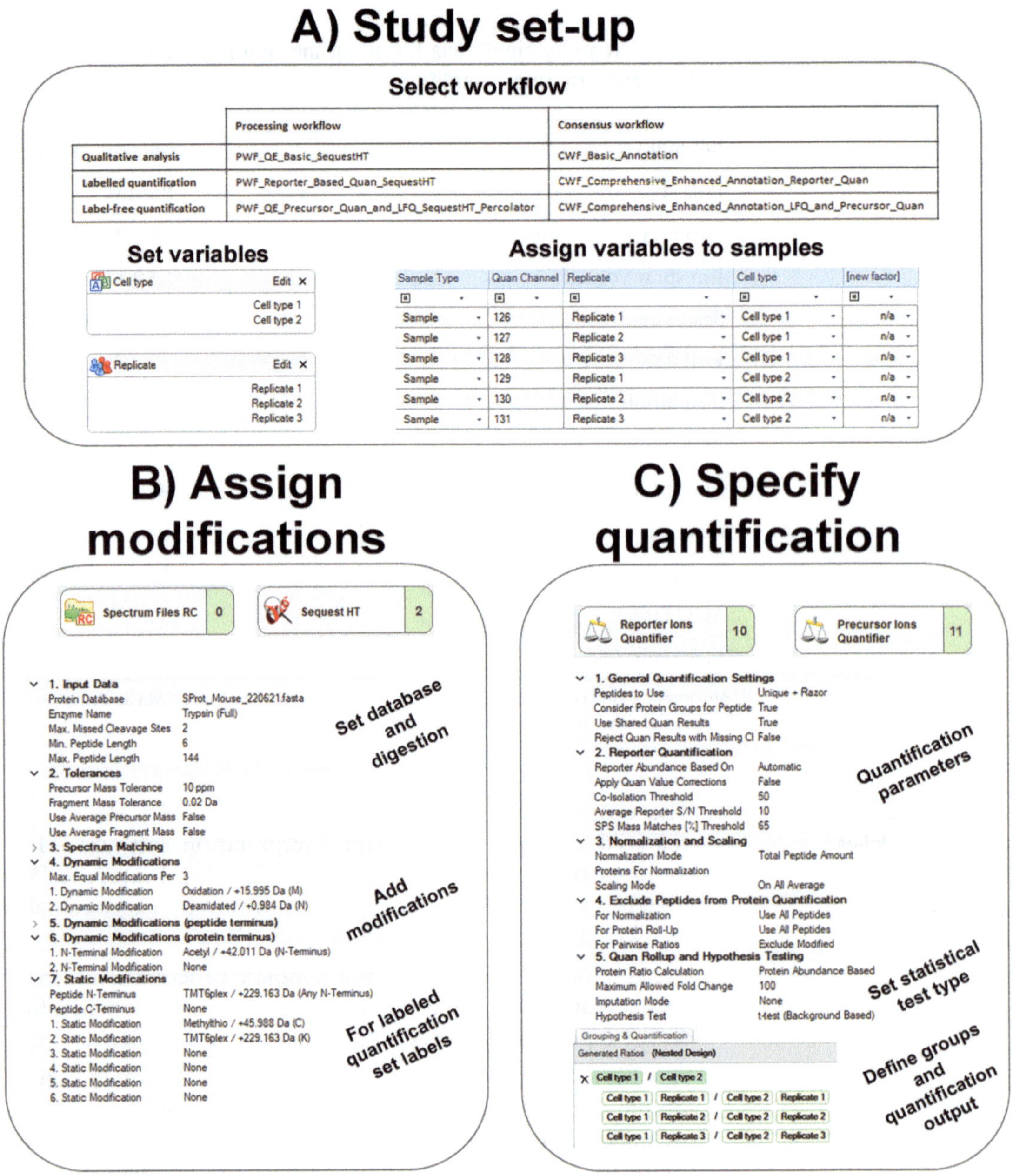

Fig. 3 Example parameters for analysis of mass spectra using Proteome Discoverer. (**a**) The study setup is used to define what type of analysis will be performed, set experimental variables such as replicates and categorical factors, and assign these variables to the imported samples. (**b**) For all analyses, modifications should be included in the processing workflow. (**c**) The reporter ion quantification (for labelled quantification) and precursor ion quantification (for label-free quantification) nodes include parameters for quantification analysis. The grouping and quantification tab can be used to define which groups should be compared to generate protein ratios

acetylation, and alkylation should be included (Fig. 3b) (*see* **Note 5**). Additionally, the predicted protein database from the species of interest and the enzyme used for protein digestion needs to be included. For labelled quantification experiments such as iTRAQ and TMT, the labels should be included as static modifications.

5. The default false discovery rates (FDR) are set to 0.01 (1%) for high and 0.05 (5%) for relaxed (*see* **Note 6**).
6. For reporter ion quantification (labelled) and precursor ion quantification (label-free), the quantification parameters should be set in their respective nodes (Fig. 3c). The type of statistical test used depends on the experimental design. In the grouping and quantification parameters, the grouping variables and the desired ratio outputs are set. For labelled quantification, the label channel can be used as a grouping variable.

3.8 Refinement of Proteome Data for Downstream Analyses

1. For both qualitative and quantitative analysis, the data output should be refined to remove low-confidence results. This applies across all proteomics software. Here we describe data refinement for the Proteome Discoverer software.
2. Firstly, it is important to ensure that you only consider "master proteins" in your post-data processing analysis. Master proteins indicate that the identified protein has significant matches to peptides that are unique to that particular protein. "Candidate master proteins" and "non-master proteins" have only achieved peptide matches with peptides that are common between multiple proteins; thus, the confidence of these proteins is lower. This may be termed differently in different software.
3. The protein FDR should also be considered as it informs on protein confidence. As stated above, we would typically set this at high (FDR <1%), medium (FDR <5%), and low (FDR >5%). Therefore, you may consider only using medium and high-confidence proteins and discarding any proteins that are only supported by low-confidence peptides. This parameter is typically applied across all spectral matching software.
4. An additional measure that can help scrutinize the resulting output is the number of peptides identified per protein. Ideally, a minimum of two peptides is preferred for each protein; however, single peptide protein matches can also be considered for identification purposes [17, 25].

3.9 Qualitative Mass Spectrometry Data

Once the results have been refined, the resulting list gives the protein content of the sample. Protein identification can provide invaluable information on many aspects of oomycete biochemistry. Functional assignment of proteins can be used to better understand the protein content, and if this information is not available for the organism of interest, functional assignment, including Gene Ontology, PFAM or Interpro, can be assigned using various tools (*see* **Note 7**). As an extension, enrichment of Gene ontologies can be obtained from a sample compared to the assignment of the proteins of the entire genome [17, 31].

3.10 Analysis of Quantitative Mass Spectrometry Data

The experimentation between label-free and labelled proteomics differs by the addition of isobaric tags on the peptides in the labelled workflow and that all samples to be compared are combined after labelling and for LC-MS analysis. These approaches have advantages and limitations to their application (*see* Table 1); however, the analysis output is typically very similar.

1. Quantitative analysis generates ratios, p-values, and abundances of each identified protein between different samples. For labelled quantification, additional information is provided for the label as a modification on each peptide.
2. Proteins that have significant p-values ($p < 0.05$) can be used to determine differentially abundant proteins between samples. The ratio of these significant proteins details the fold change of the protein and therefore informs on whether the protein abundance is significantly higher or lower.

3.11 Gene Modelling Using Peptide Data: Proteogenomics

Proteomics has traditionally been used as a tool to profile proteomes [1]. However, for proteomics to be effective, it requires a well-annotated genome and an accurate comprehensive predicted protein dataset. Improvement in sequencing technology and cost has resulted in an increase in genome sequences. Traditionally RNAseq and gene-calling software have been used to generate these predicted protein datasets [10]. Recently, several studies have exploited the use of proteomic datasets to annotate genomes and genes to improve protein datasets [23, 24]. Here, we describe an alternate application for proteomic data for genome annotation which uses a whole genome 6-frame translation and mass spectra to improve the predicted proteome [23]. This concept is termed "proteogenomics" [10].

1. In addition to matching mass spectra to the genome annotation of the organism of interest through Proteome Discoverer, they can also be matched to the 6-frame translation. This can be performed using the 6-frame translation as the database in Proteome Discoverer.
2. The resulting 6-frame peptides can then be mapped back to their genomic location (i.e., generation of a .gff3 file for peptide spectral matches) (Fig. 4).
3. BEDtools and CDS Mapper are subsequently used on this peptide.gff3 dataset alongside the genome annotation .gff3 file to determine the proximity and frame matching of spectral matches to currently annotated genes [24, 32].

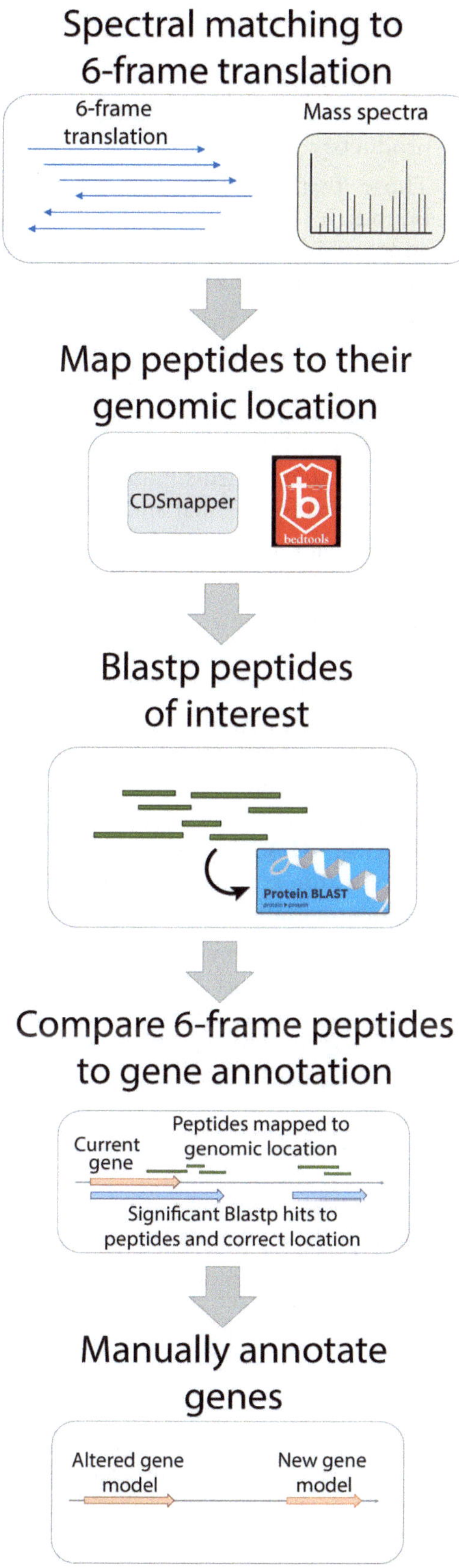

Fig. 4 A proteogenomic pipeline for the manual curation of gene models using spectral data. Mass spectra are matched to the 6-frame translation of the genome. These 6-frame peptides are mapped to their genomic

4. For the 6-frame peptides that coincide with current annotations, this provides additional validation for these gene products.
5. Those that have conflicting localization such as spanning boundaries of exons or are located where there are no currently annotated genes provide the opportunity for alteration, editing, and manual curation of new unsuspected gene models.
6. For the peptides that challenge current gene models, BLASTp can be used to determine if these peptides show homology to known proteins in species within the same genus. BLASTp parameters should be set as follows: expect threshold: $2E^5$, matrix: PAM30, gap costs: existence 9, and extension 1. Only peptides that return significant E scores ($e < 1E^{-3}$) should be considered for manual annotation [23].
7. Additionally, tBLASTn can be used to determine whether these novel peptides show homology to predicted genes.
8. In Geneious Prime (or alternatives such as UGENE), import the assembly scaffold, the current gene models, and the 6-frame peptides of interest.
9. The Blast function can be used to transfer the significant BLASTp and tBLASTn hits onto the genome. If they coincide with the current peptide location and comply with sequence features such as start/stop codons and unsequenced regions, then the BLASTp protein hits can be used to manually edit current annotations and curate new gene models. Be conscious of the integration of introns and frameshifts here.
10. Once a set of newly curated annotations has been created, additional support can be gained if the new sequences have a functional assignment (i.e., Protein Families) or can be further validated by other methods such as transcriptomics and functional studies [33].

4 Notes

1. Here we have defined the use of a Thermo Ultimate 3000 RS liquid chromatography system coupled to a QE-HF Exactive mass spectrometer and associated trap and analytical columns. This analysis is not restricted to this specific system. Various

Fig. 4 (continued) location. These peptides are compared to currently annotated genes and BLASTp is used for the peptides of interest to determine their homology to known genes. If these homologous genes comply with sequence features, they can be used to guide the manual alteration and annotation of new genes

other LC systems can be used for peptide elution and many other mass spectrometers (not restricted to fusion orbitraps) have the capability to perform this kind of analysis.

2. Each of these examples has varying capabilities as to which types of analysis can be performed. For example, Protein Pilot cannot perform unlabelled quantification. Check software specifications to ensure it can perform the analysis required. However, the examples given here are user-friendly. There are several other analysis tools available that can perform these tasks, but they may require more specialized skills.
3. For labelling, the pH of the buffer used to resuspend the sample may affect the labelling efficiency. Ensure the recommended buffers are used.
4. Ideally, reverse phase high pH fractionation should be applied when performing a labelled experiment. Fractions should be concatenated into 12 fractions for mass spectrometry analysis. This separates peptides based on hydrophobicity and enables more depth of proteome [34, 35]. Fractionation cannot be used for unlabelled experiments as the peptide quantification relies on matching retention times of the mass spectra.
5. Across the processing and consensus workflows, several nodes require the type of mass analyzer. For the QE-HF Exactive, FTMS should be selected. However, if a different type of analyzer has been used, this parameter must be updated; otherwise, the analysis will not work.
6. The FDR reflects on peptide and protein confidence. An FDR of 1% means that the confidence of the peptide is ≥99%. An FDR of 5% means that the peptide confidence is ≥95%.
7. Functional assignment of proteins by their Gene Ontologies, Interpro Scan identifications, Kyoto Encyclopedia of Genes and Genomes or Protein Families can be assigned using tools such as Interpro Scan or EGGNOG-Mapper.

References

1. Aslam B, Basit M, Nisar MA et al (2017) Proteomics: technologies and their applications. J Chromatogr Sci 55:182–196. https://doi.org/10.1093/chromsci/bmw167
2. Shepherd SJ, van West P, Gow NA (2003) Proteomic analysis of asexual development of *Phytophthora palmivora*. Mycol Res 107:395–400. https://doi.org/10.1017/S0953756203007561
3. Pang Z, Srivastava V, Liu X, Bulone V (2016) Quantitative proteomics links metabolic pathways to specific developmental stages of the plant pathogenic oomycete *Phytophthora capsici*. Mol Plant Pathol. https://doi.org/10.1111/mpp.12406
4. Resjö S, Ali A, Meijer HJG et al (2014) Quantitative label-free phosphoproteomics of six different life stages of the late blight pathogen *Phytophthora infestans* reveals abundant phosphorylation of members of the CRN effector family. J Proteome Res 13:1848–1859. https://doi.org/10.1021/pr4009095
5. Hosseini S, Resjö S, Liu Y et al (2015) Comparative proteomic analysis of hyphae and germinating cysts of *Phytophthora pisi* and *Phytophthora sojae*. J Proteome 117:24–40.

https://doi.org/10.1016/j.jprot.2015.01.006

6. Grenville-Briggs LJ, Avrova AO, Hay RJ et al (2010) Identification of appressorial and mycelial cell wall proteins and a survey of the membrane proteome of *Phytophthora infestans*. Fungal Biol 114:702–723. https://doi.org/10.1016/j.funbio.2010.06.003
7. Meijer HJG, Mancuso FM, Espadas G et al (2014) Profiling the secretome and extracellular proteome of the potato late blight pathogen *Phytophthora infestans*. Mol Cell Proteomics 13:2101–2113. https://doi.org/10.1074/mcp.M113.035873
8. Wilm M (2009) Quantitative proteomics in biological research. Proteomics 9:4590–4605. https://doi.org/10.1002/PMIC.200900299
9. Patel VJ, Thalassinos K, Slade SE et al (2009) A comparison of labeling and label-free mass spectrometry-based proteomics approaches. J Proteome Res 8:3752–3759. https://doi.org/10.1021/pr900080y
10. Renuse S, Chaerkady R, Pandey A (2011) Proteogenomics Proteomics 11:620–630. https://doi.org/10.1002/pmic.201000615
11. Thines M, Choi YJ (2016) Evolution, diversity, and taxonomy of the *Peronosporaceae*, with focus on the genus *Peronospora*. Phytopathology 106:6–18. https://doi.org/10.1094/phyto-05-15-0127-rvw
12. Le Berre JY, Engler G, Panabières F (2008) Exploration of the late stages of the tomato-*Phytophthora parasitica* interactions through histological analysis and generation of expressed sequence tags. New Phytol 177:480–492. https://doi.org/10.1111/j.1469-8137.2007.02269.x
13. Blackwell WH (2011) The genus Lagena (Stramenopila: oomycota), taxanomic history and nomenclature. Phytologia 93:157–167
14. Latijnhouwers M, De Wit PJGM, Govers F (2003) Oomycetes and fungi: similar weaponry to attack plants. Trends Microbiol 11:462–469. https://doi.org/10.1016/J.TIM.2003.08.002
15. Zhang C, Cui T, Zhang F et al (2020) Identification of differentially activated pathways in *Phytophthora sojae* at the mycelial, cyst, and oospore stages by TMT-based quantitative proteomics analysis. J Proteome 221:103776. https://doi.org/10.1016/j.jprot.2020.103776
16. Pang Z, Srivastava V, Liu X, Bulone V (2017) Quantitative proteomics links metabolic pathways to specific developmental stages of the plant-pathogenic oomycete *Phytophthora capsici*. Mol Plant Pathol 18:378–390. https://doi.org/10.1111/MPP.12406
17. Andronis CE, Jacques S, Lipscombe R, Tan K-C (2022) Comparative sub-cellular proteome analyses reveals metabolic differentiation and production of effector-like molecules in the dieback pathogen *Phytophthora cinnamomi*. J Proteome 269:104725. https://doi.org/10.1016/j.jprot.2022.104725
18. Grenville-Briggs LJ, Avrova AO, Bruce CR et al (2005) Elevated amino acid biosynthesis in *Phytophthora infestans* during appressorium formation and potato infection. Fungal Genet Biol 42:244–256. https://doi.org/10.1016/j.fgb.2004.11.009
19. Rocafort M, Fudal I, Mesarich CH (2020) Apoplastic effector proteins of plant-associated fungi and oomycetes. Curr Opin Plant Biol 56:9–19. https://doi.org/10.1016/J.PBI.2020.02.004
20. Ellis JG, Rafiqi M, Gan P et al (2009) Recent progress in discovery and functional analysis of effector proteins of fungal and oomycete plant pathogens. Curr Opin Plant Biol 12:399–405
21. Severino V, Farina A, Fleischmann F et al (2014) Molecular profiling of the *Phytophthora plurivora* secretome: a step towards understanding the cross-talk between plant pathogenic oomycetes and their hosts. PLoS One 9:e112317. https://doi.org/10.1371/journal.pone.0112317
22. Resjö S, Brus M, Ali A et al (2017) Proteomic analysis of *Phytophthora infestans* reveals the importance of cell wall proteins in pathogenicity. Mol Cell Proteomics 16:1958–1971. https://doi.org/10.1074/mcp.M116.065656
23. Andronis CE, Hane JK, Bringans S et al (2021) Gene validation and remodelling using proteogenomics of *Phytophthora cinnamomi*, the causal agent of dieback. Front Microbiol 12:1519. https://doi.org/10.3389/fmicb.2021.665396
24. Bringans S, Hane JK, Casey T et al (2009) Deep proteogenomics; high throughput gene validation by multidimensional liquid chromatography and mass spectrometry of proteins from the fungal wheat pathogen *Stagonospora nodorum*. BMC Bioinformatics 10:1–9. https://doi.org/10.1186/1471-2105-10-301
25. Casey TM, Khan JM, Bringans SD et al (2017) Analysis of reproducibility of proteome coverage and quantitation ising isobaric mass tags (iTRAQ and TMT). J Proteome Res 16:384–392. https://doi.org/10.1021/acs.jproteome.5b01154

26. Schmitthenner AF, Bhat RG (1994) Useful methods for studying *Phytophthora* in the laboratory. Ohio Agricultural Research and Development Center, Wooster, Ohio
27. Wessel D, Flügge UI (1984) A method for the quantitative recovery of protein in dilute solution in the presence of detergents and lipids. Anal Biochem 138:141–143. https://doi.org/10.1016/0003-2697(84)90782-6
28. Orsburn BC (2021) Proteome discoverer – a community enhanced data processing suite for protein informatics. Proteomes 9:15. https://doi.org/10.3390/proteomes9010015i
29. Stepanchenko NS, Novikova GV, Moshkov IE (2011) Protein quantification. Russ J Plant Physiol 58:737–742. https://doi.org/10.1134/S1021443711040182
30. Brunelle JL, Green R (2014) One-dimensional SDS-polyacrylamide gel electrophoresis (1D SDS-PAGE). Methods Enzymol 541:151–159. https://doi.org/10.1016/B978-0-12-420119-4.00012-4
31. McGowan J, O'hanlon R, Owens RA, Fitzpatrick DA (2020) Comparative genomic and proteomic analyses of three widespread *Phytophthora* species: *Phytophthora chlamydospora, Phytophthora gonapodyides and Phytophthora pseudosyringae.* Microorganisms 8:653. https://doi.org/10.3390/microorganisms8050653
32. Quinlan AR, Hall IM (2010) BEDTools: a flexible suite of utilities for comparing genomic features. Bioinformatics 26:841–842. https://doi.org/10.1093/bioinformatics/btq033
33. Ashurst JL, Collins JE (2003) Gene annotation: prediction and testing. Annu Rev Genomics Hum Genet 4:69–88. https://doi.org/10.1146/annurev.genom.4.070802.110300
34. Wang Y, Yang F, Gritsenko MA et al (2011) Reversed-phase chromatography with multiple fraction concatenation strategy for proteome profiling of human MCF10A cells. Proteomics 11:2019–2026. https://doi.org/10.1002/PMIC.201000722
35. Yang F, Shen Y, Camp DG, Smith RD (2012) High-pH reversed-phase chromatography with fraction concatenation for 2D proteomic analysis. Expert Rev Proteomics 9:129–134. https://doi.org/10.1586/epr.12.15
36. Seymour SL, Hunter CL (2015) ProteinPilot™ software overview high quality, in-depth protein identification and protein expression analysis. Available via Sciex. https://www.sciex.com/content/dam/SCIEX/pdf/posters/ProteinPilot-Software-Overview.pdf
37. Palomba A, Abbondio M, Fiorito G et al (2021) Comparative evaluation of MaxQuant and proteome discoverer MS1-based protein quantification tools. J Proteome Res 20:3497–3507. https://doi.org/10.1021/acs.jproteome.1c00143
38. Koenig T, Menze BH, Kirchner M et al (2008) Robust prediction of the MASCOT score for an improved quality assessment in mass spectrometric proteomics. J Proteome Res 7:3708–3717
39. Deutsch EW, Mendoza L, Shteynberg DD et al (2023) Trans-proteomic pipeline: robust mass spectrometry-based proteomics data analysis suite. J Proteome Res 22:615–624

Chapter 16

Heterologous Expression of Secreted Proteins from *Phytophthora* in *Pichia pastoris* Followed by Protein Purification by Immobilized Metal Ion Affinity (IMAC)

Mariana Tarallo, Max Hayhurst, Trevor S. Loo, Monica L. Gerth, and Rosie E. Bradshaw

Abstract

This chapter describes the protocol for heterologous expression of *Phytophthora* proteins in the yeast *Pichia pastoris*. Two methods to prepare the constructs for expression are described, using two different strains of *P. pastoris*, as well as methods for protein expression and purification by immobilized metal ion affinity (IMAC). The combined protocol was used to successfully produce a *Phytophthora pluvialis* small, secreted cell death elicitor protein and a *Phytophthora agathidicida* glycoside hydrolase 12, PaXEG1, also a cell death-inducing protein.

Key words Heterologous protein production, Protein purification, Immobilized metal ion affinity, Yeast transformation

1 Introduction

The yeast *Pichia pastoris* is widely used for heterologous expression of eukaryotic proteins. This expression system has the advantage of being a simple technique that generates high yields of both apoplastic and cytoplasmic proteins [1, 2] when compared to other eukaryotic expression systems (e.g., mammalian cell culture). The *P. pastoris* system also has an advantage over bacterial expression systems, as the possibility of posttranslational modifications allows for appropriate protein processing and folding [3]. The expression and purification of proteins enables, among many other things, functional characterization in molecular plant pathology studies. Here, we outline the use of two alternative expression vectors, requiring different strains of *P. pastoris*, to express proteins of interest from *Phytophthora* species, followed by protein purification using immobilized metal ion affinity (IMAC).

Monica L. Gerth and Rosie E. Bradshaw (eds.), *Phytophthora: Methods and Protocols*, Methods in Molecular Biology, vol. 2892, https://doi.org/10.1007/978-1-0716-4330-3_16,

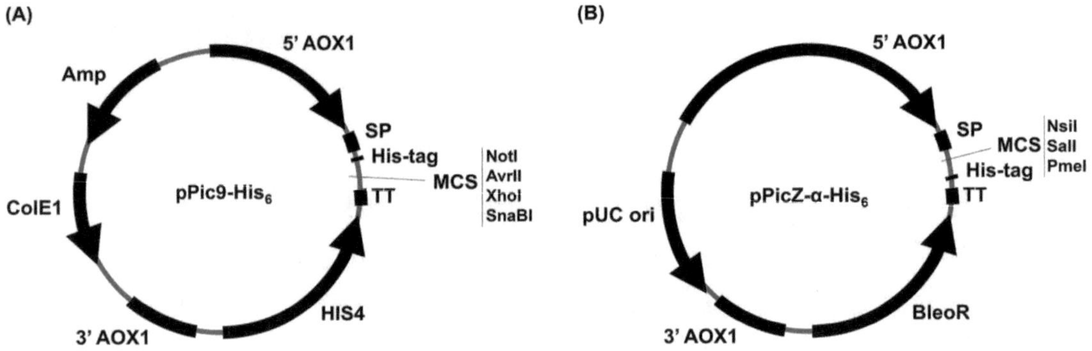

Fig. 1 Plasmid maps for the vectors used for heterologous protein expression in *Pichia pastoris*. (**a**) pPic9-His_6 and (**b**) pPicZ-α plasmid maps. *5′ AOX1*, *AOX1* gene promoter, *3′ AOX1*, *AOX1* gene terminator, *ColE1* and *pUC ori Escherichia coli* replication origin, *SP* signal peptide, *MCS* multiple cloning site, *TT* transcription terminator, *Amp* ampicillin resistance gene, *BleoR* zeocin resistance gene

A widely used expression vector is pPic9-His_6, which is used with histidine auxotrophs of *P. pastoris* such as strain GS115, which is defective in the histidine dehydrogenase gene [4]. The pPic9-His_6 plasmid (Invitrogen) (Fig. 1a) contains the *HIS4* gene, allowing *P. pastoris* transformants to be selected on a medium without histidine. The yeast transformation involves recombination of the vector into the alcohol oxidase 1 (*AOX1*) locus in the *P. pastoris* genome (the plasmid can be linearized prior to transformation to allow for such recombination). Expression from the AOX1 promoter can be induced by methanol. Thus, changing the carbon source of *P. pastoris* from glycerol to methanol will induce the expression of the gene of interest [5]. The expression vector also contains an α-factor secretion signal upstream of the multiple cloning site (MCS) and a His_6-tag. These features allow for the straightforward construction of a plasmid encoding the protein of interest with an N-terminal secretion signal and His_6-tag for purification by IMAC.

Other commonly used expression vectors for *P. pastoris* are the pPicZ-α plasmids (Invitrogen) (Fig. 1b). These have been used to express a range of different *Phytophthora* proteins [5–7]. The pPicZ-α plasmids are suitable for both restriction-ligation cloning and Gibson assembly [8]. One key distinction between pPic9 and pPicZ-α is that pPicZ-α plasmids contain the *Bleo*^{R} gene (encoding resistance to the antibiotic zeocin) as a selectable marker for transformation, rather than the *His4* marker gene. By increasing the zeocin concentration, it is possible to select for transformants with multi-copy integrations, which can lead to higher levels of heterologous protein production. Because multi-copy integrations are relatively rare, we have provided modified protocols for high-efficiency electroporation of *P. pastoris* that can increase the frequency of multi-copy transformants. Another difference between the two plasmids is the position of the His_6-tag, which, in pPicZ-α, is downstream of the MCS and, as a result, the candidate proteins contain a C-terminal His_6-tag for purification by IMAC. Choosing a vector system will, therefore, depend on the desired selection mechanism, protein yield, and His_6-tag position for each candidate protein (*see* **Note 1**).

When secreted, the recombinant protein is usually the dominant species; however, further purification can be achieved using various methods [9]. IMAC, as described in this chapter, provides a high level of selective purification based on a high-capacity ability to bind metal ions. IMAC protein purification is based on the principle that proteins containing specific amino acid residues, such as histidine and tryptophan, form complexes with the metal ions [10]. The IMAC method also has advantages over other types of chromatography in that it can be used with different matrices, such as resins, agarose, or magnetic beads, providing versatility and cost-effectiveness [11].

2 Materials

2.1 Construct Preparation Using Vector pPic9-His₆

1. pPic9-His_6 (Invitrogen).
2. Template gDNA or cDNA containing the *Phytophthora* gene to be expressed.
3. Taq polymerase, dNTPs, and buffer.
4. Appropriate restriction enzymes and buffers.
5. Gel electrophoresis apparatus.
6. Polymerase chain reaction (PCR) purification kit.
7. T4 DNA ligase and buffer.
8. Electrocompetent DH5α *Escherichia coli* cells (or similar strain suitable for cloning recombinant vectors).
9. Electroporator cuvettes (0.2 cm–gap).
10. Lysogeny broth (LB): 10 g/L tryptone, 5 g/L NaCl, 5 g/L yeast extract, 1 g/L glucose. Adjust the volume accordingly with water and add 15 g/L agar. For appropriate selection, add 100 μg/mL of ampicillin.
11. Plasmid extraction kit.
12. Primers: 5′ AOX1 (GACTGGTTCCAATTGACAAGC) and 3′ AOX1 (AATGGCATTCTGACATCC) for checking the construct.

2.2 P. pastoris *GS115 Transformation (for pPic9-His_6)*

1. *P. pastoris* GS115 strain (Invitrogen), or another strain lacking histidine dehydrogenase function.
2. 50 mL conical centrifuge tube.

3. 250 mL flask.
4. Yeast Extract–Peptone–Dextrose (YPD) medium: 10 g/L yeast extract, 20 g/L peptone, 20 g/L glucose. Adjust the volume accordingly with water and add 15 g/L agar, where necessary.
5. Sterile water.
6. 1 M sorbitol.
7. Electroporator cuvettes (0.2 cm–gap).
8. Minimal dextrose (MD) agar: 13.4 g/L yeast nitrogen base (YNB) with ammonium sulphate and without amino acids, 4×10^{-4} g/L biotin, 20 g/L glucose. Make stock solutions of all components and add to autoclaved water and 15 g/L agar.
9. 25 mM NaOH.

2.3 Construct Preparation Using Vector pPicZ-α

1. pPicZ-α-His_6 (Invitrogen).
2. Template gDNA or cDNA containing the *Phytophthora* gene to be expressed.
3. Taq polymerase, dNTPs, and buffer.
4. DpnI restriction enzyme.
5. Gel electrophoresis apparatus.
6. PCR purification kit.
7. An appropriate Gibson assembly reaction mix (e.g., NEBuilder HiFi DNA Assembly Mix or GeneArt Gibson Assembly HiFi Master Mix).
8. Electrocompetent DH5α *Escherichia coli* cells (or similar strain suitable for cloning recombinant vectors).
9. Electroporator and cuvettes (0.2 cm gap width).
10. Low-salt LB agar: 10 g/L tryptone, 5 g/L yeast extract, 5 g/L NaCl. Adjust pH to 7.5 using 5 M NaOH. Adjust the volume accordingly with water and add 15 g/L agar. For appropriate selection, add 100 μg/mL of zeocin.
11. Plasmid extraction kit.
12. Primers: 5′ AOX1 (GACTGGTTCCAATTGACAAGC) and 3′ AOX1 (AATGGCATTCTGACATCC).

2.4 P. pastoris X-33 Transformation (for pPicZ-α Plasmids)

1. *P. pastoris* X-33 strain (Invitrogen) or another strain susceptible to zeocin.
2. 4-(2-Hydroxyethyl)-1-piperazineethanesulfonic acid (HEPES) buffer: 1 M HEPES. Adjust pH to 7.5 using 5 M NaOH. Adjust the volume accordingly with water, filter sterilize, and store at room temperature (RT).
3. 1 M Dithiothreitol (DTT).

4. Electroporator cuvettes (0.2 cm gap width).
5. YPD-sorbitol (YPDS) medium: 50 g/L YPD, 1 M sorbitol. Adjust the volume accordingly with water. For selection plates, include 15 g/L agar. When cooled to below 50 °C, supplement agar with 100 μg/mL of zeocin or, to isolate multi-copy clones, with 500–2000 μg/mL zeocin.

2.5 Protein Expression (for Both Vector Types)

1. YPD medium (*see* Subheading 2.2).
2. Buffered complex glycerol medium (BMGY): 10 g/L yeast extract, 20 g/L mycological peptone, 100 mM potassium phosphate buffer (pH 6.0), 13.4 g/L YNB, 0.4 mg/L biotin, 1% (v/v) glycerol. Dissolve the yeast extract and mycological peptone in distilled water (dH_2O) and autoclave it. Filter-sterilized solutions of the remaining ingredients should then be mixed into the broth at room temperature.
3. Buffered complex methanol medium (BMMY): As BMGY, but glycerol is substituted for 0.5% (v/v) methanol.
4. 60% (v/v) glycerol, sterilized.
5. Methanol.
6. Miracloth, autoclaved.
7. 50 mL conical centrifuge tubes.
8. 125 mL and 1000 mL flasks.
9. Westen blot apparatus.

2.6 Protein Purification

1. Resin: Ni Sepharose™ 6 Fast Flow (Cytiva); *see* **Note 2** for other options.
2. Column: Glass Econo-Column (BioRad) or equivalent (2.5 times larger than resin bed volume) with column adaptor.
3. Chromatography System: NGC™ Quest/ Discover (BioRad); *see* **Note 3** for other options.
4. Büchner flask and magnetic stirrer.
5. Binding buffer: 20 mM Na_2HPO_4, 0.5 M NaCl, pH 7.4.
6. Elution buffer: 20 mM Na_2HPO_4, 0.5 M NaCl, 0.5 M imidazole, pH 7.4.
7. Syringe and syringe filter of 0.22 μm pore size.
8. 1 M NaOH.
9. Sintered glass funnel.
10. Packing reservoir.
11. Vacuum system.

3 Methods

3.1 Construct Preparation Using pPic9-His$_6$

Both the construct preparation for *P. pastoris* transformation and the *P. pastoris* transformation using pPic9-His$_6$ are based on the methods of Kombrink [4].

1. Determine the sequence encoding the signal peptide of your gene of interest, allowing the design of primers to amplify only the sequence that encodes the mature protein. The forward primer should contain the sequence for the SmaI restriction site, followed by the FLAG-tag sequence (5′ CCCGACTACAAGGACGACGATGACAAG 3′) for western blot detection, and then the gene-specific sequence (approximately 15 nucleotides). The reverse primer should contain the restriction site for EcoRI. See **Note 4** about the choice of restriction sites and the addition of a His$_6$-tag if a different plasmid is used.
2. Using PCR, amplify the gene of interest with the primers designed above. Visualize fragments by gel electrophoresis. If a single product is obtained, purify the PCR product using a commercial PCR purification kit. If multiple products are observed, extract the amplicon of the correct size from the gel and purify using a commercial gel purification kit.
3. Digest both the purified PCR product and pPic9-His$_6$ plasmid with the appropriate restriction enzymes in a final reaction volume of 20 μL.
4. After digestion, precipitate the DNA (from the 20 μL reaction) by adding 125 μL of 100% ethanol and 5 μL of 3 M sodium acetate (pH 5.2). Incubate at −20 °C overnight. Centrifuge at 17,000× *g* for 30 min at 4 °C. Discard the supernatant and wash the pellet with 125 μL of 70% ice-cold ethanol. Centrifuge at 17,000× *g* for 10 min at 4 °C. Discard the supernatant, let the pellet air-dry and resuspend it in 30–50 μL of water (or TE buffer).
5. Ligate the purified digested PCR product and pPic9-His$_6$ vector together using T4 ligase in a total volume of 20 μL.
6. Use 5 μL of the ligation reaction to transform electrocompetent DH5α *E. coli* cells, applying a charging voltage of 1500 V and resistance of 200 Ω. Immediately after electroporation add 1 mL of LB to the cuvette and transfer cells to a 1.5 mL microcentrifuge tube. Incubate cells at 37 °C with 180 rpm shaking for 1 h.
7. Spread the transformed cells across LB agar plates supplemented with ampicillin (100 μg/mL) and incubate overnight at 37 °C.

8. Select positive colonies that contain the recombinant vector by performing a colony PCR using the forward primer AOX1 (GACTGGTTCCAATTGACAAGC) and the specific candidate gene reverse primer. To do this, use a sterile pipette tip to slightly touch the colonies and transfer cells to a 200 μL microcentrifuge tube for the PCR. Positive colonies should produce a PCR product, while negative colonies will not.
9. Inoculate positive colonies in 4 mL of LB supplemented with ampicillin (100 μg/mL) and incubate overnight at 37 °C with 180 rpm shaking.
10. Extract plasmids from the cells using a commercial plasmid extraction kit.
11. After extraction, use AOX1 forward and reverse primers (*see* Subheading 2.1) to sequence and check the expression cassette.
12. Linearize the recombinant expression vector with the restriction enzyme SacI. Precipitate DNA as described in **step 3**. Between 2 and 15 μg of DNA is needed for *P. pastoris* transformation. *See* **Note 5** about the choice of restriction enzyme.

3.2 Construct Preparation Using pPicZ-α: Gibson Assembly Method

The gene of interest can be cloned into the MCS between the N-terminal α-factor secretion signal and C-terminal His_6-tag similar to pPic9 (*see* Subheading 3.1). Alternatively, inserts can be cloned into the vector by Gibson assembly [8], which we have described here.

1. Determine the coding sequence for the gene of interest. Using a primer design tool such as NEBuilder Assembly Tool or SnapGene, design Gibson assembly PCR primers to amplify both the pPicZ-α plasmid and gene of interest. The primers should add 15–30 bp homology arms to the ends of both PCR products, which enables assembly later. Gibson assembly does not rely on restriction sites, so the plasmid PCR primers can be designed to omit the MCS sequences encoding linker amino acids. If desired, include the FLAG-tag sequence in the forward primer (*see* Subheading 3.1, **step 1**).
2. In separate PCRs, amplify the pPicZ-α plasmid and gene of interest and purify using a commercial PCR purification kit.
3. Carry out a restriction digest with the plasmid PCR product using DpnI to remove the template plasmid (*see* **Note 6**). Purify the DNA with a commercial PCR purification kit.
4. Using both PCR products, carry out a Gibson assembly reaction with an appropriate product (see materials Subheading 2.3). Prepare the reaction according to the manufacturer's guidelines.

5. Use 2 μL of the Gibson assembly reaction mixture to transform electrocompetent DH5α *E. coli* cells by applying a charging voltage of 1500 V and resistance of 200 Ω. It is also prudent to include a negative control using only DpnI-digested PCR product (*see* **Note 7**). Immediately after electroporation add 1 mL of low-salt LB to the cuvette and transfer cells to a 1.5 mL microcentrifuge tube. Incubate cells at 37 °C with 180 rpm shaking for 1 h.
6. Spread the transformed cells across low-salt LB agar plates supplemented with zeocin (100 μg/mL) and incubate overnight at 37 °C. Zeocin is sensitive to pH and salt concentration and is not active when used with LB medium.
7. Select positive colonies by colony PCR (*see* Subheading 3.1, **step 8**).
8. Inoculate positive colonies in 4 mL of low-salt LB supplemented with zeocin (100 μg/mL) and incubate overnight at 37 °C with 180 rpm shaking.
9. Extract, sequence, and linearize plasmids as described in Subheading 3.1 (**steps 10–12**). It is best to sequence plasmids from at least three colonies; in our experience, Gibson assembly frequently introduces mutations near the intersection of the insert and vector.

3.3 P. pastoris Transformation

P. pastoris transformation with recombinant plasmids can be done with a standard electrocompetent cells protocol for both GS115 and X-33 strains. However, a high efficiency electrocompetent cells protocol can be performed to transform *P. pastoris* X-33 with recombinant plasmid to increase the proportion of multi-copy transformants.

1. To generate *P. pastoris* electrocompetent cells, inoculate a single colony of *P. pastoris* into 5 mL of YPD and incubate overnight at 28 °C with 250 rpm shaking.
2. Use this 5 mL of grown culture to inoculate 75 mL of YPD in a 250 mL flask. Incubate for approximately 16 h at 28 °C with 250 rpm shaking, until it reaches an OD_{600} of 1.3–1.5.
3. For standard electrocompetent cells, proceed from **step** 7. For high-efficiency electrocompetent cells, continue from **step 4**.
4. Harvest *P. pastoris* cells by centrifuging for 5 min at 4 °C at 1500× *g*. Discard supernatant and resuspend cells in 16 mL of sterile water.
5. Add 4 mL of 1 M HEPES buffer (pH 7.5) and incubate cell suspension at 30 °C for 45 min with 100 rpm shaking.
6. After 45 min of incubation, add 500 μL of 1 M DTT and incubate cell suspension at 30 °C for 15 min with 100 rpm shaking.

7. For all the following steps, keep all flasks on ice.
8. Centrifuge the cell suspension for 5 min at 4 °C at 1500× *g*. Discard the supernatant and resuspend cells in 50 mL of ice-cold water.
9. Repeat **step 8**, but resuspend cells in 25 mL of ice-cold water.
10. Repeat **step 8**, but resuspend cells in 10 mL of ice-cold 1 M sorbitol.
11. Repeat **step 8**, but resuspend cells in 10 mL of ice-cold 1 M sorbitol.
12. Repeat **step 8**, but resuspend cells in 1 mL of ice-cold 1 M sorbitol.
13. Transform 80 μL electrocompetent cells by adding 1–15 μg of linearized DNA (Subheading 3.1). For high-efficiency electrocompetent cells, 1–2 μg of linearized DNA is usually sufficient. Incubate on ice for 5 min in an electroporation cuvette. The volumes described will allow for 12 transformations (*see* **Note 8**).
14. Place the cuvette in an electroporator set and apply a charging voltage of 1500 V and resistance of 200 Ω.
15. For standard electrocompetent cells, follow **steps 16** and **17**, or for high-efficiency electroporation, skip to **steps 18** and **19**. For both methods, finish with colony PCR (**step 20**).
16. Immediately after electroporation add 1 mL of ice-cold 1 M sorbitol to the cuvette and transfer cells to a 1.5 mL microcentrifuge tube.
17. Plate 200 μL of the cell suspension on MD plates. Centrifuge the remaining cells for 4 min at 6000× *g* at room temperature and resuspend the pellet in 80–100 μL of 1 M sorbitol, plating them on MD plates. Incubate plates at 28 °C for 2–3 d.
18. Immediately after electroporation, add 600 μL ice-cold YPDS medium and transfer cell suspension to a 15 mL tube. Incubate at 30 °C, 200 rpm for 1 h to recover.
19. Plate recovered cells on YPDS plates with zeocin ranging from 100 μg/mL to 2000 μg/mL and incubate at 30 °C for 2–3 d (*see* **Note 9**).
20. Select positive colonies by performing a colony PCR using forward and reverse primers for the specific candidate gene. To do this, use a sterile pipette tip to transfer cells from each colony to a 200 μL microcentrifuge tube containing 30 μL of 25 mM of NaOH. Incubate tubes for 10 min at 99 °C and use 1 μL of the mixture to perform the colony PCR.
21. Purify positive transformants by streak-plating each positive transformant on YPD plates.

3.4 Protein Expression

This protocol for protein expression in *Pichia pastoris* is based on that of Weidner et al. [12], with minor modifications.

Because considerable variation can be found between independent *P. pastoris* transformants, it is important to determine which transformants express the protein of interest and their optimal time for expression. Normally, screening three to five independent transformants should be enough to identify at least one that expresses the protein of interest.

1. For each purified *P. pastoris* transformant to be analyzed, inoculate one single colony into 6 mL of BMGY in a sterile 50 mL centrifuge tube. Grow culture overnight at 28 °C with 250 rpm shaking until the OD_{600} reaches 2 (*see* **Note 10**).
2. Once the culture reaches the right OD_{600}, transfer 1 mL to a 2.0 mL microcentrifuge tube and add 1 mL of 60% (v/v) sterile glycerol. Freeze the glycerol stock and store at −80 °C.
3. Centrifuge the remaining culture at 3000× *g* for 5 min at room temperature. Discard the supernatant and remove any excess media with a 1 mL pipette. Residual glycerol from BMGY media can inhibit protein expression.
4. Resuspend the pellet in 4 mL of BMMY and repeat the centrifugation step. Discard the supernatant and resuspend the pellet in 20 mL of BMMY.
5. Transfer the resuspended culture to sterile 125 mL flasks and cover them with two layers of miracloth (or nappy liner). Incubate the flasks at 28 °C with 250 rpm shaking (*see* **Note 11**).
6. Add 0.5% (v/v) of methanol to each flask every 24 h to induce protein expression.
7. Collect samples (approximately 1 mL) 24, 48, and 72 h after induction for western blot analysis [13].
8. After the best *P. pastoris* clones and expression conditions have been selected, induce the expression of candidate proteins for the appropriate time. Proportionally adjust the volume of BMGY and BMMY depending on the volume of CE protein required. For example, we inoculated 25 mL of BMGY; cells were resuspended in 20 mL of BMMY, and 100 mL were used to induce expression in a 1 L flask (*see* **Note 11**).
9. Collect cells by centrifugation at 4500× *g* for 15 min at 4 °C.
10. Transfer the supernatant, containing the secreted protein, to a fresh tube and repeat the centrifugation **step 9**.
11. To remove any remaining *P. pastoris* cells, sterilize the supernatant by passing it through a 0.22 μm membrane filter.
12. Before the protein purification step, adjust the pH of the supernatant to pH 8 by the addition of 1 M NaOH. Sometimes

a precipitate forms after pH adjustment; in this case pass the supernatant through another 0.22 μm membrane filter before protein purification.

3.5 Protein Purification

Before performing the purification, it is best to test the culture filtrate from *P. pastoris* containing the empty expression vector to check it is a valid background control. For example, if you are studying candidate effector proteins to determine if they trigger a cell-death type response in plant tissue, it is important to ensure that the empty vector filtrate does not elicit a cell-death response in the absence of the protein. If it does, this suggests that the *P. pastoris* secreted something else into the culture filtrate that might be recognized by the plant immune system (e.g., a microbe-associate molecular pattern) and, therefore, the purification step should be performed.

This protocol outlines how to purify a secreted candidate protein (fused with an N-terminal His_6-tag) from *P. pastoris* culture filtrate by immobilized metal ion chromatography using a nickel-charged resin. The resin is packed into a column, and the sample is passed through the column to achieve separation. The sample can be passed through the column by gravity flow (small scale), centrifugation (spin), or automated pump-based system for larger sample volumes, which was used here. See **Notes 2** and **3** for alternatives to the automated system.

3.5.1 Packing the Column

This is a general procedure that suits most resins. Refer to the manufacturer's instructions for column characteristics about resin binding capacity, recommended flow rate, operating pressure, and chemical/pH stability.

1. Pour the appropriate amount of well-suspended resin into a sintered glass funnel and wash with 10 column volumes of water (column volume is the volume of resin inside a column) to remove the alcohol preservative under vacuum. The resin used here (Ni Sepharose™ 6 Fast Flow (Cytiva)) has a suggested protein binding capacity of 40 mg/mL, so 1–3 mL of resin would be enough for protein purification from a 100 mL culture.
2. Equilibrate the resin with 5 column volumes of binding buffer and transfer to a Büchner flask for degassing under vacuum for 15 min with gentle mixing on a magnetic stirrer at room temperature.
3. Gently pour the degassed gel slurry into a column with the outlet closed. Allow the gel to settle for several minutes and open the outlet to facilitate packing. A short and wide-diameter column is recommended to reduce column back pressure and mass transfer resistance, for improved resolution, and rapid purification.

4. If available, attach a column adaptor and adjust until the plunger is just above the resin bed. Pump buffer through the column at twice the desired flow rate or at the maximum pressure allowed by the column hardware and resin for several minutes to compact the resin. Readjust the position of the column adapter until it just touches the settled gel bed. Column adaptors reduce the head space above the column, allowing efficient delivery of a buffer gradient and preventing uneven flow, peak broadening, and a loss of resolution.

3.5.2 Purification on a FPLC System

Refer to the manufacturer's instructions for FPLC operation.

1. Purge the entire flow path of the chromatography system with water then with prepared buffers according to manufacturer's instructions.
2. Attach the packed column to the FPLC system and wash with 10 column volumes of binding buffer to condition the column. Metal ion leakage during purification can be minimized by performing a blank run before loading sample: Wash with 5 column volumes of elution buffer, followed by 10 column volumes of binding buffer.
3. Load the sample onto the IMAC column at one-third to half the desired flow rate to allow extra time for binding which may improve yield. Collect the flow-through in fractions appropriate to the size of the column (usually in fractions of 1 column volume). For larger columns, reduce the size of the fractions to 0.2 to 0.5 column volumes, depending on the collection vessel used.
4. Wash the column with binding buffer until the absorbance reaches the baseline or at least 10 column volumes have been passed through.
5. Elute the protein with elution buffer using a linear gradient or in multiple steps of increasing percentage of elution/binding buffer mix (e.g., 0–50% elution buffer over 10–20 column volumes in a linear gradient, or five steps in 10% elution buffer increments for 5 column volumes each). Optimize the elution strategy after this test run, either extend or change the concentration of elution buffer in wash steps according to the elution profile. Include a wash step containing an imidazole concentration slightly lower than what is necessary to elute the target protein. This will increase purity by removing the majority of unbound contaminants, without eluting the bound His-tagged protein. The final elution should contain a concentration of imidazole greater than the calculated concentration corresponding to the eluted peak of the target protein (*see* **Note 12**).

6. Save an aliquot (approximately 50–100 μL) for western blot detection to check the size and integrity of the candidate protein [13].
7. It is recommended to use purified proteins right after the purification step. However, the solution containing the protein can be stored at −20 °C, but it is not recommended to freeze and thaw the solution multiple times.

4 Notes

1. Both plasmids described have some advantages and disadvantages. The pPic9-His_6 plasmid involves the use of a histidine auxotroph for the selection of transformants and is a well-established system; this is cost-effective and sufficient for most proteins and applications. The pPicZ-α plasmids rely on antibiotic selection, which has two main benefits. Firstly, the pPicZ-α plasmids are suitable for use in any *P. pastoris* strains susceptible to the antibiotic, offering more potential for protein expression to be optimized by strain choice. Secondly, the selection pressure on the yeast can be modified by changing the antibiotic concentration. By using higher concentrations of zeocin, it is possible to select for transformants with multi-copy plasmid integrations, which is not possible with pPic9. In theory, multi-copy transformants should express larger quantities of protein, which can be advantageous for large-scale experiments, or for proteins with a low yield. Finally, the His_6-tag is located at a different site in each plasmid: the N-terminus in pPic9, and the C-terminus in pPicZ-α. The presence and location of a His_6-tag can affect the expression and properties of recombinant proteins [14, 15] so it may even be desirable to test both plasmids in parallel.
2. Alternatives for protein purification without an automated system:

   Pre-packed columns—gravity-flow based:

   - His GraviTrap™ (Cytiva), nickel-based.
   - His60 Ni Gravity Columns (Takara), nickel-based.
   - HisTALON™ Gravity (Takara), cobalt-based.

   Pre-packed spin columns:

   - HIS-Select® Spin (Millipore), nickel charged.
   - His-Spin Protein Miniprep (Zymo Research), nickel-based.
   - HisPur™ Cobalt Spin/Ni-NTA Spin (Thermo Fisher Scientific), comes with cobalt or nickel.
   - His SpinTrap (Cytiva), nickel charged.

- Ni-NTA Spin (Qiagen), nickel-based.
- TALON® Spin (Takara), cobalt-based.

3. Alternatives for protein purification with an automated system:
   - ÄKTA™ start/prime plus/pure/avant (Cytiva).
   - AZURA® Bio Lab/Bio Pilot (Knauer).
4. If SmaI and/or EcoRI are present in the candidate gene, use alternative restriction sites shown in Fig. 1a, ensuring that the gene will be ligated into the plasmid in the correct orientation so that the promoter (5′ AOX1) is upstream. If an alternative plasmid is used that does not contain the His_6-tag, add the sequence (CATCATCATCATCATCAT) between the SmaI restriction site and the FLAG-tag sequence.
5. If a SacI recognition site is present in the candidate gene, alternative sites include NsiI, SalI, and PmeI (Fig. 1b). For more information, refer to the Invitrogen Instruction manual for plasmid pPic9-His6.
6. DpnI is a restriction enzyme that digests methylated DNA. Since PCR products are not methylated, DpnI can be used to specifically destroy template plasmid prior to transformation. For this to work, the plasmid must be purified from a *dam* methylation competent (dam^+) strain of *E. coli* (DH5α is suitable).
7. The template-only control will indicate whether the DpnI digest removed enough of the original pPicZ-α plasmid. If there is a large number of colonies on control plates, try increasing the DpnI digest time, re-purifying the plasmid PCR product, or using fresh reagents.
8. *P. pastoris* electrocompetent cells should be prepared on the day of the transformation; it is not advised to store these cells. Transformants can still be obtained from electrocompetent cells stored at −80 °C. However, the transformation efficiency drops significantly.
9. Using the pPicZ-α plasmid, this protocol generally obtains transformation efficiencies of approximately 4×10^5 CFU per μg of plasmid using 100 μg/mL zeocin selection and 3×10^3 using 1000 μg/mL zeocin selection. We usually plate 10 μL of cells on 100 μg/mL zeocin, and the rest (100 μL per plate) on YPDS plates with 1000 μg/mL or 2000 μg/mL zeocin.
10. Perform a 20-fold dilution of samples for a more accurate result when measuring the OD_{600} on the spectrophotometer.
11. Do not use 50 mL centrifuge tubes during this step, as they will not provide the right aeration for protein expression. Baffled flasks can be used to introduce more oxygen. For protein induction, the culture volume should not be higher than 10–30% of the total flask volume.

12. Bound proteins can be eluted by lowering the buffer pH. The histidine residues become protonated and are unable to bind to the immobilized ion. Weakly bound proteins can be eluted readily at pH 6.0, while strongly bound proteins can be eluted successively when pH is lowered from 6.0 to 4.0. Typical buffers used are 20 mM sodium acetate, 0.5 M NaCl, pH 6, and pH 4. To avoid protein denaturation of the fractions with prolonged contact at low pH, pipette 1 M Tris–HCl, pH 8.0 (100–200 μL/ml fractions) to quickly restore the pH to neutral. Do not use NaOH. It is recommended to test the sensitivity of the target protein at low pH ranges [16].

References

1. Zheng J, Guo N, Lin F-L, Wu L-S, Zhou H-B (2014) Screening of multi-copy mannanase recombinants of *Pichia pastoris* based on colony size. World J Microbiol Biotechnol 30 (2):579–584. https://doi.org/10.1007/s11274-013-1479-x
2. Werten MWT, Eggink G, Cohen Stuart MA, de Wolf FA (2019) Production of protein-based polymers in *Pichia pastoris*. Biotechnol Adv 37(5):642–666. https://doi.org/10.1016/j.biotechadv.2019.03.012
3. Karbalaei M, Rezaee SA, Farsiani H (2020) *Pichia pastoris*: a highly successful expression system for optimal synthesis of heterologous proteins. J Cell Physiol 235(9):5867–5881. https://doi.org/10.1002/jcp.29583
4. Kombrink A (2012) Heterologous production of fungal effectors in *Pichia pastoris*. In: Bolton MD, Thomma BPHJ (eds) Plant fungal pathogens: methods and protocols. Humana Press, pp 209–217. https://doi.org/10.1007/978-1-61779-501-5_13
5. Larroque M, Ramirez D, Lafitte C, Borderies G, Dumas B, Gaulin E (2011) Expression and purification of a biologically active *Phytophthora parasitica* cellulose binding elicitor lectin in *Pichia pastoris*. Protein Expr Purif 80 (2):217–223. https://doi.org/10.1016/j.pep.2011.07.007
6. Wu C-H, Yan H-Z, Liu L-F, Liou R-F (2008) Functional characterization of a gene family encoding polygalacturonases in *Phytophthora parasitica*. Mol Plant-Microbe Interact 21 (4):480–489. https://doi.org/10.1094/MPMI-21-4-0480
7. Xia Y, Ma Z, Qiu M, Guo B, Zhang Q, Jiang H, Zhang B, Lin Y, Xuan M, Sun L, Shu H, Xiao J, Ye W, Wang Y, Wang Y, Dong S, Tyler BM, Wang Y (2020) N-glycosylation shields *Phytophthora sojae* apoplastic effector PsXEG1 from a specific host aspartic protease. Proc Natl Acad Sci 117(44):27685–27693. https://doi.org/10.1073/pnas.2012149117
8. Gibson DG, Young L, Chuang R-Y, Venter JC, Hutchison CA, Smith HO (2009) Enzymatic assembly of DNA molecules up to several hundred kilobases. Nat Methods 6(5):343–345. https://doi.org/10.1038/nmeth.1318
9. Liu S, Li Z, Yu B, Wang S, Shen Y, Cong H (2020) Recent advances on protein separation and purification methods. Adv Colloid Interf Sci 284:102254. https://doi.org/10.1016/j.cis.2020.102254
10. Rodriguez EL, Poddar S, Iftekhar S, Suh K, Woolfork AG, Ovbude S, Pekarek A, Walters M, Lott S, Hage DS (2020) Affinity chromatography: a review of trends and developments over the past 50 years. J Chromatogr B 1157:122332. https://doi.org/10.1016/j.jchromb.2020.122332
11. Cheung RCF, Wong JH, Ng TB (2012) Immobilized metal ion affinity chromatography: a review on its applications. Appl Microbiol Biotechnol 96(6):1411–1420. https://doi.org/10.1007/s00253-012-4507-0
12. Weidner M, Taupp M, Hallam SJ (2010) Expression of recombinant proteins in the methylotrophic yeast *Pichia pastoris*. J Vis Exp 36:1862. https://doi.org/10.3791/1862
13. Kurien BT (2021) Basics of Western blotting. In: Kurien BT (ed) Western blotting for the non-expert. Springer International Publishing, pp 7–21. https://doi.org/10.1007/978-3-030-70684-5_2
14. Sabaty M, Grosse S, Adryanczyk G, Boiry S, Biaso F, Arnoux P, Pignol D (2013) Detrimental effect of the 6 his C-terminal tag on YedY enzymatic activity and influence of the TAT signal sequence on YedY synthesis. BMC Biochem 14(1):28. https://doi.org/10.1186/1471-2091-14-28

15. Phan TTP, Tran LT, Schumann W, Nguyen HD (2015) Development of Pgrac100-based expression vectors allowing high protein production levels in *Bacillus subtilis* and relatively low basal expression *in Escherichia coli*. Microb Cell Factories 14(1):72. https://doi.org/10.1186/s12934-015-0255-z

16. Hemdan ES, Zhao YJ, Sulkowski E, Porath J (1989) Surface topography of histidine residues: a facile probe by immobilized metal ion affinity chromatography. Proc Natl Acad Sci 86 (6):1811–1815. https://doi.org/10.1073/pnas.86.6.1811

Index

Monica L. Gerth and Rosie E. Bradshaw (eds.), *Phytophthora: Methods and Protocols*, Methods in Molecular Biology, vol. 2892, https://doi.org/10.1007/978-1-0716-4330-3,

Q

R

S

T

Y

MIX
Papier aus verantwortungsvollen Quellen
Paper from responsible sources
FSC® C105338

If you have any concerns about our products,
you can contact us on
ProductSafety@springernature.com

In case Publisher is established outside the EU,
the EU authorized representative is:
Springer Nature Customer Service Center GmbH
Europaplatz 3, 69115 Heidelberg, Germany

Printed by Libri Plureos GmbH
in Hamburg, Germany